Between **Science** and **Economics** – Vol. 2

Quantum Computing

Physics, Blockchains, and Deep Learning Smart Networks

Melanie Swan
Purdue University, USA

Renato P. dos Santos
Lutheran University of Brazil, Brazil

Frank Witte
University College London, UK

World Scientific

NEW JERSEY • LONDON • SINGAPORE • BEIJING • SHANGHAI • HONG KONG • TAIPEI • CHENNAI • TOKYO

Published by

World Scientific Publishing Europe Ltd.
57 Shelton Street, Covent Garden, London WC2H 9HE
Head office: 5 Toh Tuck Link, Singapore 596224
USA office: 27 Warren Street, Suite 401-402, Hackensack, NJ 07601

Library of Congress Control Number: 2019053514

British Library Cataloguing-in-Publication Data
A catalogue record for this book is available from the British Library.

Between Science and Economics — Vol. 2
QUANTUM COMPUTING
Physics, Blockchains, and Deep Learning Smart Networks

ISBN 978-1-78634-820-3 (hardcover)
ISBN 978-1-78634-821-0 (ebook for institutions)
ISBN 978-1-78634-822-7 (ebook for individuals)

For any available supplementary material, please visit
https://www.worldscientific.com/worldscibooks/10.1142/Q0243#t=suppl

Desk Editors: Balasubramanian/Shi Ying Koe

Typeset by Stallion Press
Email: enquiries@stallionpress.com

About the Authors

Melanie Swan is a research associate at the UCL Centre for Blockchain Technologies, a technology theorist in the Philosophy Department at Purdue University, and a Singularity University faculty member. She is the founder of several startups including the Institute for Blockchain Studies, DIYgenomics, GroupPurchase, and the MS Futures Group. Melanie's educational background includes an MBA in Finance from the Wharton School of the University of Pennsylvania, a PhD in Philosophy from Purdue University, and a BA in French and Economics from Georgetown University. She is the author of the best-selling book *Blockchain*: *Blueprint for a New Economy*. Other notable work is on the topics of blockchain economics, human brain–cloud interface, BCI cloudminds, multigenic risk assessment, the Brain as a DAC (Decentralized Autonomous Corporation), neural payment channels, biocryptoeconomics, and block time (the native time domain of blockchains).

Renato P. dos Santos is a researcher on blockchain technologies and Graduate Professor at the Lutheran University of Brazil. He is a member of the British Blockchain Association, holds a DSc (Physics) degree and did post-doc works in artificial intelligence, and specializations in data science and blockchain technologies. He is also the author of more than 100 scientific papers about philosophy of cryptocurrencies, data science in STEM education, second life in STEM education, Web 2.0 technologies, ethnoscience, physics teaching, artificial intelligence and computer algebra in physics, and quantum field theory in prestigious scientific periodicals and events around the world. He is a reviewer and editor of prestigious scientific periodicals and events around the world and developed systems for second life, forex market, qualitative physics, and computer algebra.

Frank Witte completed an MSc in Theoretical Astrophysics (1992) at Utrecht University in the Netherlands and received his PhD in Theoretical Physics (1995) from the University of Heidelberg. As an assistant and associate professor he taught theoretical physics at Utrecht University and University College Utrecht (1997–2010) and published on diverse topics such a phase-transitions in non-equilibrium field theories, bound-states of fermions under gravitational interactions and the foundations of quantum game theory. As his research interests shifted towards the application of physics-inspired methods and concepts in economics he accepted a position in the Department of Economics of University College London where he is working today. He teaches economics of science, with a forthcoming textbook to be published by World Scientific, and environmental economics as well as computational methods for economists. Frank has spent extended academic visits, including teaching and research, at St. John's College, Cambridge (UK), the Quantum Optics & Laser Science group at Imperial College London and as International Fellow of Grinnell College (US).

Contents

List of Figures

List of Tables

Chapter 1
Introduction

Quantum computing ... has established an unprecedentedly deep link between the foundations of computer science and the foundations of physics

— John Preskill (2000, p. 127)

The implication is that reality can be computed. The startling premise of quantum computing is that the bizarre quantum mechanical realm that was thought incomprehensible can be computed. Quantum mechanics might remain incomprehensible, but at least it can be computed. The bigger notion is that the link between quantum computing and quantum physics suggests the possibility of delivering on the promise of understanding the true nature of physical reality.

Merely the first step is simulating the quantum mechanical world with quantum computers in its image as Feynman envisioned. Beyond that milestone, the real endpoint is mobilizing the quantum mechanical domain to compute more difficult problems and create new possibilities.

Engaging the quantum realm makes it clear that tools are needed to comprehend the two domains together, the microscale of the quantum and the macroscale of lived reality. The issue concerns not only understanding more about quantum mechanics, but also linking the quantum and non-quantum domains such that the quantum realm can be activated in a useful way at the macroscale. This book is a journey towards a model to do

Figure 1.1. Model of computational reality.

precisely this, incorporating the superposition, entanglement, and interference properties of quantum mechanics with the 3D time and space geometries of the macroscale world.

With the assumption that physical reality can be computed with quantum information systems, the theme of computation is taken up through the lens of smart network technologies (self-operating computation networks) and their potential expansion into the quantum mechanical domain. A technophysics approach (the application of physics principles to the study of technology) is used to derive smart network field theory (conventional and quantum) as a physical theory of smart network technologies. The organizing principle of this work is encapsulated in a causal model (Figure 1.1).

The model of computational reality posits that computation capability influences the ability to have knowledge about physical reality. Computation capability may be moderated by variables such as the properties of computational systems, the theories and algorithms that manage the computation process, and security features that protect and allow system criticality to be managed, essentially the tools and apparatus by which the capacity for efficient computation is deployed towards the end of understanding reality. The main hypothesis is that computation is the biggest current factor in acquiring knowledge about physical reality at all scales, from the observable universe to the Planck scale.

1.1 Quantum Futures?

Some speculative yet reasonable picture of the near future could unfold as follows. Quantum computing is farther along than may be realized, but is still a very early-stage technology with considerable uncertainty about its possible progression. However, given the fact that early-stage quantum computers have been demonstrated and are shipping, and that perhaps a hundred or more worldwide labs are engaged in high-profile, well-funded, competitive research efforts, it seems possible that better quantum computers may become available in the next few decades. The path is clear but non-trivial, the task is scaling up the number of qubits from 30–70 to hundreds, thousands, and millions, and testing various quantum error correction schemes for the end goal of constructing a universal fault-tolerant quantum computer.

Many kinds of quantum computing hardware are being demonstrated, and it is possible that optical quantum computing could accelerate in the same vein as optical global communications did previously, this time for the quantum internet with satellite-based quantum key distribution, secure end-to-end communications, quantum routers and modems, and distributed quantum computing. Smart network technologies such as blockchain and deep learning are sponsoring the advance to quantum computing with post-quantum cryptography and quantum machine learning optimization and simulation, and other smart network technologies could follow suit, such as autonomous vehicle networks, robotic swarms, automated supply chains, and industrial robotics cloudminds. Quantum computing is not just nice, but necessary to keep pace with growing data volumes, and to enable a new slate of applications such as whole brain emulation, atomically-precise manufacturing, and causal models for disease.

1.2 Technophysics

Technophysics is the application of physics principles to the study of technology, by analogy to Biophysics and Econophysics. Biophysics is an interdisciplinary science that uses the approaches and methods of physics to study biological systems with the goal of understanding the structure, dynamics, interactions, and functions of biological systems. Likewise, Econophysics is an interdisciplinary field of research that applies theories and methods developed in physics to solve problems in economics and finance, particularly those including uncertainty, stochastic processes, and nonlinear dynamics. Considering Technophysics, although much of technology development is initially physics-based, a wider range of concepts and methods from ongoing advances in physics could be applied to the study and development of new technology. In this work, the technophysics approach is directed at the study of smart network technologies (intelligent self-operating networks such as blockchains and deep learning neural networks), by engaging statistical physics and information theory for the characterization, control, criticality assessment, and novelty catalysis of smart network systems.

Technophysics arises from a confluence of factors in contemporary science. First, there have been several non-trivial connections in the last few decades linking the physical world and the representational world of mathematics, algorithms, and information theory. One example is that statistical physics has made a link between spin glasses in condensed matter physics and combinatorial optimization problems. Specifically, an association between the glass transition in condensed matter physics, error-correcting codes in information theory, and probability theory in computer science has been established with statistical physics (Mezard & Montanari, 2009). Another example is the symbiotic relationship of using machine learning algorithms to understand physical systems, and using analogies with physical systems and materials to understand the operation of algorithms (Krzakala *et al.*, 2007). A third example is the holographic principle (Susskind, 1995), formalized in the AdS/CFT correspondence (Maldacena, 1998), and its link to information theory (Harlow & Hayden, 2013), quantum error correction codes (Pastawski *et al.*, 2015), and tensor networks (Witten, 2016). The holographic principle suggests that in any physical system, there is a correspondence between a volume of space and its boundary region. The implication is that the interior bulk can be described by a boundary theory in one fewer dimensions, in an information compression mechanism between the 3D bulk and the 2D boundary.

A second factor giving rise to technophysics is a cross-pollination in academic methods. Various approaches have become part of the cannon of scientific methods irrespective of field. These methods include complexity science, network theory, information science, and computation graphs. Many fields now have a computational information science counterpart, both in the hard sciences (such as computational neuroscience, computational chemistry, and computational astrophysics), and in liberal arts (seen in the digital humanities, semantic data structure investigation, and text mining and analysis). The opposite trend is also true, developing a physics-based counterpart to fields of study, such as in Biophysics, Econophysics, and Technophysics. The exchange between academic methods is leading to the development of a more comprehensive and robust apparatus of universal scientific study across all areas of the academe.

1.2.1 *Conceptual toolkit of ideas*

The premise of this text is that it is necessary to be facile with technophysics ideas to be effective in today's world. Hence, a compendium of knowledge modules to enable such facility is presented in the course of this book. Having a grasp of concepts and methods from a broader range of fields that extend beyond an initial field of training, especially from contemporary research frontiers and interstices, could increase the capacity for innovation and impact in the world. This book provides a strategy for managing the considerable uncertainty of the disruptive possibilities of quantum computing in the next decades and beyond with smart network field theories as a tool. Methods are integrated from statistical and theoretical physics, information theory, and computer science. The technophysics approach is rooted in theory and application, and presented at the levels of both defining a conceptual understanding as well as outlining how formalisms and analysis techniques may be used in practice.

It is possible to elaborate some of the standardized ideas that comprise the canon of technophysics knowledge. These include eigenvalues and eigenvectors, and the notion of operators that measure the multidimensional configuration of a system and that can act on the system. The Hamiltonian is the operator par excellence that measures the total energy in a system, and allows problems to be written in the form of an energy minimization problem. Complementarity (only one property of a quantum system can be measured at a time), time dilation (the system looks different to different viewers), and geometry-based metrics are important. There is a notion of selectable parameters, among different geometries, coordinate systems, and space and time domains.

In some sense, the terms network, graph, tensor, matrix, operator, and 3D all point at the same kind of concept, and while not technically synonyms, mutually imply each other as properties of a computable system. A computation graph is a network, a network implies a computation graph or matrix, and all are 3D or of higher-order dimensionality. Further, an operator is a matrix, a field is a matrix, and tensor networks and random tensor networks can be used to rewrite high-dimensional entangled problem domains to be analytically computable. Dimensionality portability, rescaling, and reduction are frequently-used techniques. Answers are likely to be given in probabilistic terms. The bulk–boundary correspondence is a conceptual structure in which a system can be written as two regions, a surface in one fewer dimensions that can be used to interrogate the bulk region as a more complex and possibly unknowable domain.

A key technophysics principle is reducing or rendering a physical system into a form that is computable and then analytically solving it. Computational complexity is a constant focus. Classical problems can be made quantum-computable by writing them with SEI properties (superposition, entanglement, and interference), meaning converting the problem to a structure that engages the quantum properties of superposition, entanglement, and interference. As a gross heuristic, quantum computers may allow a one-tier increase in the computational complexity schema of problem calculation. Any problem that takes exponential time in classical systems (i.e. too long) may take polynomial time in quantum systems (i.e. a reasonable amount of time for practical use). In the canonical Traveling Salesman Problem, perhaps twice as many cities could be checked in half the time using a quantum computer.

1.2.2 *New slate of all-purpose smart technology features*

Smart network technologies are creating a variety of new technology developments as standard features to solve important problems, and that have greater extensibility beyond their initial purpose. One example is consensus algorithms, which provide cryptographic security of blockchains through the mining operation, and more generally comprise a standard technology (ConsensusTech) that could be used for the self-coordinated agreement and governance of any multi-agent system (human or machine). Other examples of generic technologies include zero-knowledge proofs, error correction, and hash functions, each of which, when generalized, conveys a new concept in technology.

Zero-knowledge proofs are computational proofs that are a mechanistic set of algorithms that could be easily incorporated as a feature in many technology systems to provide privacy and validation. Zero-knowledge proofs are proofs that reveal no information except the correctness of the statement (data verification is separated from the data itself, conveying zero knowledge about the underlying data, thereby keeping it private). The proofs are used first and foremost to prove validity, for example, that someone is who they claim to be. Proofs are also an information compression technique. Some amount of activity is conducted and the abstracted output is all that is necessary as the outcome (the proof evaluates to a one-bit True/False answer or some other short output). The main concept of a proof is that some underlying work is performed and a validated short answer is produced as the result.

Quantum error correction is necessary to repair quantum information bits (qubits) that become damaged, by adhering to quantum properties such as the no-cloning theorem (quantum information cannot be copied) and the no-measurement rule (quantum information cannot be measured without damaging it). Consequently, quantum error correction involves relying upon entanglement among qubits to smear out the original qubit's information onto entangled qubits which can be used to error-correct (restore) the initial qubit. The proximate use is quantum error correction. However, the great benefit is that a structural feature is created in the form of the error correction apparatus that can be more widely deployed. An error correction-type architecture can be used for novel purposes. One such project deploys the error correction feature to control local qubit interactions in an otherwise undirectable quantum annealing solver (Lechner *et al.*, 2015). The overall concept is system manipulation and control through quantum error correction-type models.

Hash functions are another example of a general-purpose smart network technology whose underlying mechanism is not new, but is finding a wider range of uses. Like proofs, hash functions are a form of information compression technology. A hash function is an algorithm that can be run over any arbitrarily large-sized digital data file (e.g. a genome, movie, software codebase, or 250-page legal contract) which results in a fixed-length code, often 32 bytes (64 alphanumeric characters). Hash functions have many current uses including password protection and the sending of secure messages across the internet such that a receiving party with the hashing algorithm and a key can decrypt the message. Hash functions are also finding novel uses. One is that since internet content can be specified with a URL, the URL can be called with a hash format by other programs (the hash function standardizes the length of the input of an arbitrarily-long URL). This concept is seen in Web 3.0 as hash-linked data structures.

The key point is the development of generic feature sets in smart network technologies that can be translated to other uses. This is not a surprise, since a property of new technology is that its full range of applications cannot be envisioned at the outset, and evolves through use. The automobile was initially conceived as a horseless carriage. What is noticeable is the theme that these features are all forms of information compression and expansion techniques (proofs and hash functions compress information and error correction expands information). This too is not surprising, given that these features are applied to information theoretic domains in which a key question is the extent to which any signal is compressible (and more generally signal-to-noise ratios and the total possible system configurations (entropy)). However, proofs and hash functions are different from traditional information compression techniques since they can convert an arbitrarily-large input to a fixed output, which connotes the attributes of a flexible and dynamical real-time system. This book (especially Chapter 15) extends these insights to interpret the emerging features (proofs, error correction, and hash functions) and quantum smart networks more generally, in a dimensional model (the bulk–boundary correspondence). In the bulk–boundary correspondence, the compression or expansion activity is performed in a higher-dimensional region (the bulk), and then translated such that the result appears in one fewer dimensions in another region (the boundary).

1.3 Chapter Highlights

This book aims to provide insight as to how quantum computing and quantum information science as a possible coming paradigm shift in computing may influence other high-impact digital transformation technologies such as blockchain and machine learning. A theoretical connection between physics and information technology is established. Smart network theory is proposed as a physical theory of network technologies that is extensible to their potential progression to quantum mechanical information systems. The purpose is to elaborate a physical basis for technology theories that is easily-deployable in the design, operation, and catalytic emergence of next-generation smart network systems. This work proposes the theoretical construct of smart network theories, specifically a smart network field theory (SNFT) and a smart network quantum field theory (SNQFT), as a foundational basis for the further development of smart network systems, and particularly quantum smart networks (smart network systems instantiated in quantum information processing environments).

There are pressing reasons for the development of smart network theories as macro-level system control theories since many smart network technologies are effectively a black box whose operations are either unknown from the outset (deep learning networks), or becoming hidden through confidential transactions (blockchain-based economic networks). Such smart networks are complex systems whose possibility for system criticality and nonlinear phase transition is unknown and possibly of high magnitude.

Towards this end, Part 1 introduces smart networks and quantum computing. Chapter 2 defines smart networks and smart network theory, and develops the smart network field theory in the classical and quantum domains. Chapter 3 provides an overview of quantum computing including the basic concepts (such as bit and qubit) and a detailed review of the different quantum hardware approaches and superconducting materials. A topic of paramount concern, when it might be possible to break existing cryptography standards with quantum computing, is addressed (estimated unlikely within 10 years, however methods are constantly improving). Chapter 4 considers advanced topics in quantum computing such as interference, entanglement, error correction, and certifiably random bits as produced by the NIST Randomness Beacon.

Part 2 provides a detailed consideration of blockchains and zero-knowledge proofs. Chapter 5 elaborates a comprehensive range of current improvements underway in classical blockchains. Chapter 6 discusses the quantum internet, quantum key distribution, the risks to blockchains, and proposals for instantiating blockchain protocols in a quantum format. Chapter 7 consists of a deep-dive into zero-knowledge proof technology and its current status and methods. Chapter 8 elaborates post-quantum cryptography and quantum proofs.

Part 3 focuses on machine learning and artificial intelligence. Chapter 9 discusses advances in classical machine learning such as adversarial learning and dark knowledge (also an information compression technique), and Chapter 10 articulates the status of quantum machine learning. The first kinds of applications being implemented in quantum computers are machine learning-related since both machine learning and quantum computation methods are applied in trying to solve the same kinds of optimization and statistical data analysis problems.

Part 4 develops the smart network field theory on the basis of statistical physics, information theory, and model field theory systems. Chapter 11 elaborates two model field theories, statistical neural field theory and spin glass theory. Chapter 12 develops the smart network field theory in detail with system elements, operation, and criticality detection measures, and considers applications in blockchain and deep learning smart network systems.

Part 5 extends the smart network field theory into the quantum realm with a model called the AdS/CFT correspondence (also known as gauge/gravity duality and the bulk–boundary correspondence) and holographic codes. Chapter 13 describes the holographic principle and its formalization in the AdS/CFT correspondence, and work connecting physical theory to information theory through the correspondence. Chapter 14 discusses the quantitative mobilization of the AdS/CFT correspondence into holographic quantum error-correcting codes.

Part 6 posits quantum smart networks as the quantum instantiation of smart networks and elaborates the smart network quantum field theory. In Chapter 15, a number of speculative conjectures are presented as to how the smart network quantum field theory may be engaged in a holographic format based on the bulk–boundary correspondence. The risks, limitations, and farther consequences of the work are discussed, proposing the possibility of multiple future eras of quantum computing.

References

Harlow, D. & Hayden, P. (2013). Quantum computation vs. firewalls. *J. High Energ. Phys*. 2013, 85.

Krzakala, F., Montanari, A., Ricci-Tersenghi, F. *et al*. (2007). Gibbs states and the set of solutions of random constraint satisfaction problems. *PNAS*. 104(25): 10318–323.

Lechner, W., Haukel, P. & Zoller, P. (2015). A quantum annealing architecture with all-to-all connectivity from local interactions. *Sci. Adv*. 1(9)e1500838.

Maldacena, J. (1998). The large N limit of superconformal field theories and supergravity. *Adv. Theor. Math. Phys*. 2:231–52.

Mezard, M. & Montanari, A. (2009). *Information, Physics, and Computation*. Oxford UK: Oxford University Press, pp. 93–168.

Pastawski, F., Yoshida, B., Harlow, D. & Preskill, J. (2015). Holographic quantum error-correcting codes: Toy models for the bulk–boundary correspondence. *J. High Energ. Phys*. 6(149):1–53.

Preskill, J. (2000). Quantum information and physics: Some future directions. *J. Modern Opt*. 47(2/3):127–37.

Susskind, L. (1995). The world as a hologram. *J. Math. Phys*. 36(11):6377–96.

Witten, E. (2016). An SYK-like model without disorder. arXiv:1610.09758 [hep-th].

Part 1

Smart Networks and Quantum Computing

Chapter 2

Smart Networks: Classical and Quantum Field Theory

Abstract

This work aims to establish a deeper connection between physics and information technology. Smart network theory is proposed as a physical theory of network technologies, particularly to encompass a potential expansion into the quantum computing domain. The objective is to elaborate a physical basis for technology theories that is easily deployable in the design, operation, and catalytic emergence of next-generation smart network systems. The general notion of smart network theory as a physical basis for smart network technologies is developed into a smart network field theory (SNFT) and a smart network quantum field theory (SNQFT) relevant to the two scale domains. The intuition is that the way to orchestrate many-particle systems from a characterization, control, criticality, and novelty emergence standpoint is through an SNFT and an SNQFT. Such theories should be able to make relevant predictions about smart network systems.

2.1 Smart Networks

Smart networks are intelligent autonomous networks, an emerging form of global computational infrastructure, in which decision-making and self-operating capabilities are built directly into the software. Examples of smart networks include blockchain economic networks, deep learning pattern recognition networks, unmanned aerial vehicles, real-time bidding (RTB) for advertising, and high-frequency trading (HFT) networks. More formally, smart networks are state machines that make probabilistic guesses about reality states of the world and act automatically based on these guesses. Communications networks are becoming computational networks in the sense of running executable code. Smart networks are a contemporary feature of reality with possibly thousands to billions of constituent elements, and thus entail a theoretically-robust model for their design and operation. Using statistical physics (statistical neural field theory and spin-glass models) and information theory (the anti-de Sitter space/conformal field theory, AdS/CFT, correspondence), this work proposes SNFTs for the orchestration of the fleet-many items in smart network systems.

2.2 Smart Network Theory

2.2.1 *Conventional (SNFT) and (SNQFT)*

There is an urgent motivation for the development of smart network theories. Smart network technologies are being demonstrated as domains of complexity (exhibiting the behavior of complex systems), which are challenging to understand and manage. Smart networks are like quantum many-body systems in which interactions become too complex to model directly. Many smart network technologies are effectively a black box whose operations are either unknown from the outset (deep learning networks), or becoming hidden through zero-knowledge proofs (blockchain economic networks). Simultaneously, smart network technologies are experiencing rapid worldwide adoption, becoming unwieldy in scale, and possibly migrating to quantum computers.

The research intuition is that a technophysics approach (the application of physics principles to the technology domain) is warranted for the further development of smart network technologies. The smart network theories proposed in this work, the smart network field theory (SNFT) and the smart network quantum field theory (SNQFT), are designed to provide an integrated theoretical basis for smart network technologies that are rooted in physical foundations. The smart network theories can be used to orchestrate many-particle systems from a characterization, control, criticality, and novelty emergence perspective.

2.2.2 *Smart network technologies are quantum-ready*

Smart network technologies are already instantiated in 3D formats (computation graphs which imply programmability and analytic solvability), which makes them somewhat quantum-ready for the next steps towards deployment in quantum computers. The 3D format suggests the possibility of instantiation in quantum information systems with superposition and multi-dimensional spaces. One next step for implementation in quantum computers would be writing smart network technologies in the form of quantum-computable properties, namely superposition, entanglement, interference (SEI properties are the quantum computable properties).

Smart network technologies are quantum-ready, meaning 3D, as a by-product of their being instantiated in a computational graph format which is 3D. This is part of a more general shift to 3D in both the computational domain and the end user domain. Technological interfaces are accommodating more 3D interactions with reality (which is itself 3D). Emblematic of 3D interfaces is that point cloud data is a new kind of internet traffic. Point cloud data captures 3D positioning information about entities (humans, robots, objects) in the context of their surroundings, with simultaneous localization and mapping (SLAM) technology. Another example of 3D-type interfaces is an advance in machine learning called geometric deep learning in which information can be analyzed in its native form rather than being compressed into lower-dimensional representations for processing. The point for quantum-readiness is that many smart network technologies such as blockchain and deep learning, as well as many other contemporary analytical systems, are already instantiated in computation graphs which are by definition 3D (and really n-dimensional), which could facilitate their potential transition to quantum information systems.

2.3 Two Eras of Network Computing

The notion of smart networks is configured in the conceptualization of there being two fundamental eras of network computing (Figure 2.1).

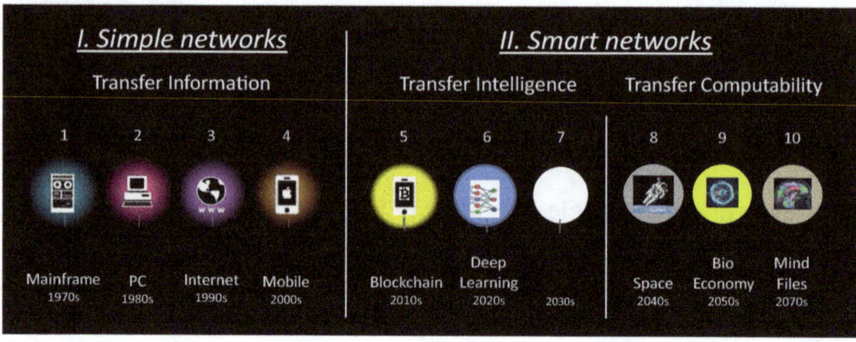

Figure 2.1. Eras of network computing: Simple networks and smart networks.

Table 2.1. Early smart networks (Smart Networks 1.0).

Internet of trading	**Internet of advertising**
High-frequency trading	Real-time bidding
MarketTech	*AdTech*
Internet of energy	**Internet of things**
Smart grids (power)	Smart city
EnergyTech	*SensorTech*

Most of the progress to date, from mainframes to mobile computing, has concerned the transfer of basic information on simple networks. Now, however, in a second phase of network computing, a new paradigm is being inaugurated, smart networks (Swan, 2015).

2.3.1 *Smart Networks 1.0*

There are different developmental phases of smart networks. An early class of smart networks (Smart Networks 1.0) can be identified (Table 2.1). Smart Networks 1.0 include HFT networks, the RTB market for advertising, smart energy grids with automated load-rebalancing, and smart city Internet of Things (IoT) sensor ecologies. The concept of smart network technologies emerged with programmatic or HFT systems for automated stock market trading. In 2016, HFT was estimated to comprise 10–40% of the total US trading volume in equities, and 10–15% of the trading volume in foreign exchange and commodities (Aldridge & Krawciw, 2017).

Another early example of a smart network technology is the RTB market, an automated marketplace for online display advertising (web-based ads). Impressions are sold in a RTB model in which advertisers bid for impressions in real time, as consumers visit websites. Although RTB is highly efficient, two pricing models persist in this market, both RTB and ahead-of-time reservation contracts (Sayedi, 2018). In other early smart network technologies, smart energy grids conduct automated load-rebalancing, in which the emergence of complex behavior has been noted, in particular the synchronization of coupled oscillators (Dorfler *et al.*, 2013). Smart city IoT sensor ecologies indicate the substantial smart network challenges that are faced in coordinating the 50 billion connected objects that are estimated to be deployed by 2020 (Hammi *et al.*, 2017).

2.3.2 *Smart Networks 2.0*

The contemporary generation of smart network technologies is Smart Networks 2.0 (Table 2.2). Although blockchain distributed ledgers and deep learning systems are some of the most prominent examples of smart network technologies, there are many kinds of such intelligent

Table 2.2. Robust self-operating smart networks (Smart Networks 2.0).

Internet of value	**Internet of analytics**
Blockchains	Deep learning
EconTech, GovTech, PrivacyTech, ProofTech	*IDTech*
Internet of goods and services	**Internet of vehicles**
Automated supply chain	Autonomous driving networks
TradeTech	*TransportTech*
Internet of brains	**Internet of qubits**
Cloudminds	Quantum internet
Medical nanorobots (BCI)	
NeuralTech	*QuantumTech*
Internet of data structures	**Internet of virtual reality**
Web 3.0	Gaming
DataTech (HashTech)	*VirtualRealityTech*

self-operating networks. Other examples include automated supply chain and logistics networks (TradeTech), autonomous vehicle networks (TransportTech), industrial robotics cloudminds, the potential quantum internet, Web 3.0's internet of data structures, and virtual reality video gaming. Smart networks operate at scales ranging from the very large, in space logistics platforms (Supply Chain 4.0) (Chen & Ho, 2018), to the very small, for control systems in brain–computer interfaces (Swan, 2016) and human brain–cloud interfaces (Martins *et al.*, 2019).

2.3.2.1 *Blockchains*

A blockchain is a distributed data structure which is cryptographically protected against modification, malicious or otherwise. Blockchains (technically, transaction blocks cryptographically linked together) are one topology among others in the more general class of distributed ledger technologies (Tasca & Tessone, 2019). Distributed ledger technology is EconTech and GovTech in the sense that institutional functions may be outsourced to computing networks (the administrative functions that orchestrate the patterns of human activity). Blockchains provide an alternative legal jurisdiction for the coordination of large groups of transnational actors using game theoretic incentives instead of policing, and economics as a design principle. Blockchains may be evolving into a new era that of PrivacyTech and ProofTech through zero-knowledge proof technology and verifiable computing.

2.3.2.2 *Machine learning: Deep learning neural networks*

Machine learning is an artificial intelligence technology comprising algorithms that perform tasks by relying on information patterns and inference instead of explicit instructions. Deep learning neural networks are the latest incarnation of artificial intelligence, which is using computers to do cognitive work (physical or mental) that usually requires a human. Deep learning neural networks are mechanistic systems that "learn" by modeling high-level abstractions in data and cycling through trial-and-error guesses with feedback to establish a system that can make accurate predictions about new data. Machine learning systems are IDtech (identification technology), which conveys the ability to recognize objects (physical or digital), by analogy to FinTech, RegTech, TradeTech, and HealthTech as standard technologies that digitize, standardize, and automate their respective domains. Objects include patterns, structures, and other topological features that are within the scope of geometrical deep learning.

The premise of deep learning is that reality comprises patterns, which are detectable through data science methods. Deep learning is notable as a smart network technology that replaces hard-coded software with a *capacity*, in the form of a learning network that is trained to perform an activity. Whereas initially software meant fixed programs running in closed domains (Software 1.0), software is starting to mean programs that dynamically engage with reality in a scope which is not fully prespecified at the outset (Software 2.0).

2.3.2.3 *Internet of data structures (Web 3.0)*

Web 3.0 means adding more functionality, collaboration, and trust to the internet (Web 1.0 was the read web, Web 2.0 the read/write web, and Web 3.0 the read/write/trust web). The idea is to install trust mechanisms such as privacy and verifiability from the beginning, directly into the software. Blockchains are in the lead of incorporating such PrivacyTech and ProofTech, and the functionality could spread to other domains.

Web 3.0 further connotes the idea of an internet of data structures. There are many different internet-based data structures such as blockchains, software codebases (Github), and various other content libraries. There may be a URL path to each content element. By using a Merkle tree structure tool, internet-based content trees become available to be called in other applications as distributed authenticated hash-linked data structures. A hash code (or simply hash) is the fixed-length output (often 32 bytes (64 characters) in blockchain protocols) of a hash function which is used to map data of arbitrary size onto data of a fixed size. A Merkle tree or hash tree is a tree in which every leaf node is labeled with the hash of a data block, and every non-leaf node is labeled with the cryptographic hash of the labels of its child nodes. Hash trees are widely used for the secure and efficient verification of the contents of large data structures.

Distributed authenticated hash-linked data structures can be deployed with a project from Protocol Labs called InterPlanetary Linked Data (IPLD). An earlier project InterPlanetary File System (IPFS) is a content-addressable file system (an improvement to file name-addressable systems which can result in errors when information paths no longer exist). IPLD is a data model for the content-addressable web in which all hash-linked data structures are treated as subsets of a unified information space, integrating all data models that link data with hashes as instances of IPLD. The web effectively becomes a Merkle forest of Merkle trees that can all be linked with interoperability through multi-hash protocols (connecting different hashing structures and content trees). These kinds of innovations are emblematic of infrastructural upgrades to the internet that facilitate privacy and security as important properties of smart network technologies.

2.3.3 *Smart Networks 3.0: Quantum smart networks*

In the farther future, quantum smart networks could comprise a next-generation of smart networks, Smart Networks 3.0 (Table 2.3). The first quantum smart network application is the quantum internet, which is already in the early stages of development for quantum key distribution and secure end-to-end communications. Quantum blockchains are a possibility, with quantum key distribution and a more substantial implementation of blockchain protocols in quantum information systems, possibly using new concepts such as proof-of-entanglement, holographic consensus, and quantum channels as the analog of payment channels. Quantum machine learning is already progressing (through quantum annealing optimization, quantum simulation, and geometric deep learning). Finally, there could be quantum brain–computer interfaces (for example, using interference-based amplitudes as a firing threshold mechanism).

Table 2.3. Quantum smart networks (Future-class Smart Networks 3.0).

Quantum internet (Q-internet)	Quantum blockchains (QBC)
Quantum machine learning networks (QML)	Quantum brain–computer interfaces (QBCI)

2.3.4 *Smart network convergence*

2.3.4.1 *Autonomous vehicle networks and robotic swarms*

Many smart networks have fleet-many items that are autonomously coordinated. This includes unmanned vehicles (aerial, underwater, and space-based), autonomous-driving vehicles (automobiles, commercial trucks, and small transportation pods), drones, robotic swarms, and industrial robots. (Robotic swarms are multiple robots that are coordinated as one system.) Autonomous vehicle networks and robotic swarms must coordinate group behavior between themselves such as flocking, foraging, and navigation, in order to carry out tasks.

Complexity theory is used to study autonomous vehicle networks because they can exhibit undesirable emergent behavior such as thrashing, resource-starving, and phase change (Singh *et al.*, 2017). Constraining flocking to optimized formations is a research focus for unmanned aerial vehicles, particularly those with autonomous strike capability (Vasarhelyi *et al.*, 2018). Optimal control theory has been proposed as one basis for the development of risk management standards in self-controlling software (Kokar *et al.*, 1999).

Applications for smart network fleets of self-coordinating machines arise in an increasing range of industrial and military use cases related to targeted material delivery, precision agriculture, and space-based and underwater exploration. Autonomous driving is a smart network technology with considerable interest. Some of the topics in contemporary research focus on collision avoidance (Woerner *et al.*, 2019) and defining a set of international standards for the levels of driving automation (SAE, 2018).

There is a convergence story in that autonomous vehicle networks and robotic swarms are using other smart network technologies such as blockchain, machine learning, and IoT sensors. Blockchains offer a variety of functionality to autonomous vehicle networks including automated record-keeping, liability-tracking, compliance-monitoring, privacy, secure communications, and self-coordination. Notably, blockchain consensus algorithms are a method by which a group of agents can reach agreement on a particular state of affairs. Consensus is a generic technology that could be used for the self-coordination of any multi-agent system, not necessarily restricted to the context of mining and transaction confirmation. Hence, the blockchain-based self-governance of robotic swarm systems has been proposed (Ferrer, 2017).

The blockchain record-logging functionality together with smart contracts could be used to automatically file insurance claims when needed, and also register discovery claims (claiming rights over discovered objects) in new venues such as undersea and space-based exploration. Smart city IoT sensor networks could be used in conjunction with blockchains and robotic swarms for consumers and businesses to request robotic swarm-as-a-service functionality to send out a swarm to conduct a sensing project. Robotic sensor swarms could survey the aftermath of accidents to summon emergency medical and police services, be engaged to provide security, and scan pipelines and other infrastructure for routine maintenance. The IoT robotic swarm model is in some sense a realization of the science-fiction idea of fleets of entrepreneur-owned mobile video cams for hire (Brin, 2002), as a sort of next-generation citizen journalism.

Verifiable computing and zero-knowledge proofs enable a new level of smart network self-coordination and control. Advanced applications, particularly for military use, could include scenarios in which agents can cooperatively work towards a solution while having minimal information. For example, the mission information could be stored in a Merkle tree such that swarm operators can see the general blueprint of the mission without the raw data being disclosed (Ferrer *et al.*, 2019). The swarm agents could use the secure communications and consensus-reaching properties of blockchains to coordinate, self-govern, and problem-solve. Further, zero-knowledge technology (which separates data verification from the underlying data) could be used in two ways, for an agent to obtain the Merkle tree-stored data relevant to its own activity, and to prove its integrity to peers by exchanging cryptographic proofs.

Various features of blockchains are implicated in this advanced swarm robotics model. The basic features are privacy and secure communication. Then, consensus technology is used for reaching a self-orchestrated group agreement without a centralized authority. Merkle tree path-addressing is used that only exposes need-to-know information. Finally, zero-knowledge proofs are used to prove earnest participation to peers without revealing any underlying personal information.

2.3.4.2 *Deep learning chains*

Deep learning chains refers to the concept of a further convergence of smart networks in the notion of a generalized control technology that has properties of both blockchain and deep learning (Swan, 2018). Deep learning chains instantiate the secure automation, audit-log tracking, remunerability, and validated transaction execution of blockchains, and the object identification (IDtech), pattern recognition, and optimization technology of deep learning. Deep learning chains in the form of blockchain-based reinforcement learning have been proposed for an air traffic control system (Duong *et al.*, 2019). Also, deep learning chains might be used as a general control technology for fleet-many internet-connected smart network technologies such as UAVs, drones, automated supply chain networks, robotic swarms, autonomous vehicle networks, and space logistics platforms. The minimal functionality of deep learning chains in autonomous driving fleets is identifying objects in a driving field (deep learning) and tracking vehicle activity (blockchain). Deep learning chains could likewise apply to the body, as a smart network control technology for medical nanorobots, identifying pathogens (deep learning) and tracking and expunging them (blockchain smart contracts). There could be greater convergence between individual smart network technology platforms (listed in Table 2.4 per their operating focus). For example, blockchains are starting to appear more regularly in the context of smart city power grid management (Pieroni *et al.*, 2018).

2.3.4.3 *Deep learning proofs*

Computational proofs are a mechanistic set of algorithms that could be incorporated as a feature in many smart network technology systems to provide privacy and validation. The potentially wide-scale adoption of zero-knowledge proof technology in blockchains makes blockchains a PrivacyTech and a ProofTech. Zero-knowledge proof technology could be similarly adopted in other smart network systems such as machine learning, for example in the idea of deep learning proofs. The first reason is the usual use case for proofs, to prove validity. This could be an important functionality in image recognition networks in autonomous driving for

Table 2.4. Smart networks by operational focus.

Smart network		Smart network operational focus
1.	Unmanned aerial vehicles (UAVs)	UAV drones with autonomous strike capability
2.	High-frequency trading (HFT)	Algorithmic trading (40% US equities), auto-hedging
3.	Real-time bidding (RTB)	Automated digital advertising placement
4.	Energy smart grids	Power grid load-balancing and transfer
5.	Blockchain economic networks	Transaction validation, self-governance, smart contracts
6.	Deep learning networks	Object identification (IDtech), pattern recognition, optimization
7.	Smart City IoT sensor landscapes	Traffic navigation, data climate, global information feeds
8.	Industrial robotics cloudminds	Industrial coordination (cloud-connected smart machines)
9.	Supply chain logistics nets	Automated sourcing, ordering, shipping, receiving, payment
10.	Personal robotic assistant nets	Personalization, backup, software updates, fleet coordination
11.	Space: aerial logistics rings	*In situ* resource provisioning, asynchronous communication

example, where the agent (the vehicle) is able to prove that certain behaviors were taken. Another reason to use proof technology is because proofs are an efficient mechanism with wider applicability beyond the proof execution context.

A central challenge in deep learning systems, which occupies a significant portion of research effort, is developing systems to efficiently calculate the error contribution of each node to the overall system processing. Various statistical error assessment methods are employed such as mean squared error (MSE), sum of squared errors of prediction (SSE), cross-entropy (softmax), and softplus (a smoothing function). An improved error contribution calculation method would be very helpful.

Proofs might be a useful solution because they are an information compression technique. Some portion of activity is conducted and the abstracted output is all that is necessary as a result (the proof evaluates to a one-bit True/False answer or some other short answer). With a proof structure, deep learning perceptrons could communicate their results using fewer information bits than they do now. The perceptron is a two-tier information system, with meta-attributes about its processing (error contribution, weights, biases) and the underlying values computed in the processing. The proof structure could be instantiated in the TensorFlow software architecture so that the proofs would be automatically generated as a feature that flows through the system's matrix multiplications. The concept of a proof is that some underlying work is performed and a validated short answer is produced as the result. The idea of deep learning proofs is that in a deep learning system, perceptrons could execute a proof of their node's contribution.

Deep learning consensus algorithms is another idea, in which consensus algorithms would be employed in deep learning systems such that perceptrons self-coordinate answers. Through the deep learning consensus algorithms, the perceptrons could self-orchestrate the processing of the nodes, and also their initial setup into an optimal configuration of layers and nodes for the problem at hand. Consensus technology is a mechanism for self-organization and governance in multi-agent systems. Deep learning consensus algorithms build on the idea of deploying consensus technologies in robotic swarms to self-coordinate to achieve mission objectives (Ferrer *et al.*, 2019).

2.4 Smart Network Field Theory: Classical and Quantum

The notion of smart network theory as a physical basis for smart network technologies is developed into the SNFT and the SNQFT, with respect to the two scale domains. The intuition is that the way to orchestrate many-particle systems from a characterization, control, criticality, and novelty emergence standpoint is through field theories such as an SNFT and an SNQFT. Such theories should be able to make relevant predictions about smart network systems as part of their operation.

Large-scale networks are a feature of contemporary reality. Such network entities are complex systems comprising thousands to billions of elements, and require a SNFT or other similar mechanism for the automated characterization, monitoring, and control of their activity. A theoretically-grounded model is needed, and smart network theories based on statistical physics (statistical neural field theory and spin-glass models), information theory (the AdS/CFT correspondence), and model systems are proposed. Generically, a *SNFT* (conventional or quantum) is a field theory for the characterization, monitoring, and control of smart network systems, particularly for criticality detection and fleet-many item management.

2.4.1 *Theory requirements: Characterize, monitor, and control*

The purpose of SNFTs is the characterization, monitoring, and control of smart network systems. The first objective is characterization. It is necessary to develop standard indicators and metrics to easily identify specific behaviors in smart network systems as they evolve and possibly grow in scalability and deployment. Both positive (emergent innovation) and negative (flash crash) behaviors should be assessable by the theory.

The second objective of an SNFT is to provide monitoring, at both the individual element and overall system level, of current and evolving behavior. Monitoring pertains to smart network operations that are currently unfolding, and also may be developing in the future. For example, in the farther future, deep thinkers (advanced deep learning systems) might go online. Although deep learning networks are currently isolated and restricted to certain computational infrastructures, it is imaginable that learning algorithms might be introduced to the internet. Risk management is a key concern. A *Deep Thinkers Registry* could be a safeguard for tracking an entity's activity, with possible annual review by a Computational Ethics Review Board for continued licensing. This is a future example that demonstrates the intended extensibility of SNFTs, and the uncertain future situations that they might help to facilitate.

The third objective of an SNFT is control, for the coordination of fleet-many items. Orchestrating fleet-many items is a clear automation economy use case for smart network technologies. This involves the ability to securely coordinate fleet-many items in any kind of internet-connected smart network system, which could include autonomous vehicles, drones, blockchain peer-to-peer nodes, deep learning perceptrons, smart city IoT sensors, home-based social robots, medical nanorobots, and supply chain shipment-receiving. The longer-term range of deployment of smart network technologies could extend to the very small, such as the cellular domains of the body, and the very large, such as civilization building in space.

2.4.1.1 *Fleet-many coordination and system criticality*

Whereas the practical application of SNFT is the automated coordination of fleet-many items, the risk management application is being able to detect and possibly avert unwanted critical moments such as potential phase transition. A crucial aspect of an SNFT is the predictive risk management of system criticality. It is important to have a mathematical and theoretical basis for understanding smart networks so that the critical points and phase transitions may be predictively managed to the extent possible. The events that could constitute criticality and phase transition in smart networks are both expected and emergent situations from both within and outside the network. Some examples are financial contagion, network security, novel technology emergence, and electromagnetic pulses.

SNFTs could also be useful in the well-formed design of smart network systems. They provide a formal scientific basis for studying smart networks as new technological objects in the contemporary world, particularly since smart networks are a nascent, evolving, and high-impact situation.

2.5 Smart Network Field Theory Development

More precisely, an SNFT is any formal method for the characterization, monitoring, control of smart network systems such as blockchains and deep learning networks. Although there are different kinds of smart networks, blockchain and deep learning are the focus for developing a field theory because they are the most sophisticated, robust, and conceptually novel.

2.5.1 *The "field" in field theory*

The term "field" is meant both analogically and literally (in the physical sense). Other terms invoked in this work such as temperature and pressure also may have both precise analytical meanings in the physical context and conceptual meanings. Terms may be applied conceptually as to the purpose and function they are serving in smart network systems.

There are two primary meanings of field in the conceptual sense. First and most generally, field refers to the ability to control multiple items as one unit. The requisite functionality of the SNFT is to manage fleet-many items. One idea is to control them as a field, in which the term field might be dynamically-defined based on location, energy, probability, gradients, or other parameters. The concept of field might be used to coordinate thousands and millions of constituent elements (such as blockchain peer-to-peer nodes or deep learning perceptrons). An example of an existing smart network field operation is optogenetics (in which neurons express a protein that makes their electrical activity controllable by light) (Boyden, 2015). Optogenetics is a "light switch for a field of neurons" in that it conveys the ability to turn on or off a field of neurons all at once. Thus, an SNFT is created, in that optogenetically enabled cells are controlled as a field as opposed to individually (Swan, 2021).

The second meaning of field in SNFTs is that a field might refer to the situation in which each element in a system has its own measure and contribution to an overall metric or network activity (possibly being used to calculate a Hamiltonian or other composite measure of a system). This concept of field (from effective field theory development in physics) suggests that every point in a landscape has a computable value, generally referring to the idea that a function has a value everywhere throughout the space, at every location in the field.

2.5.1.1 *Scalar, vector, and tensor*

SNFTs may be structured in scalar, vector, and tensor terms. A tensor is a complex mathematical object (a geometric object that maps functions and interacts with other mathematical objects). The simplest tensors are scalars (zero-rank tensors) and vectors (first-rank tensors). A scalar is a tensor with magnitude but no direction (a zero-rank point), and is described by one number. Mass and temperature are scalars. A vector is a tensor with magnitude and direction (representable by a first-rank tensor or vector line). A vector is often represented as an arrow, and defined with respect to a coordinate system. Force and velocity are vectors. A tensor is a multi-dimensional structure, representable by a matrix. Tensors are seen in the context of deep learning. Google's tensor processing units (TPUs) and TensorFlow software use tensors in the sense conducting very-fast matrix multiplications. A tensor is representable by a multi-dimensional matrix, and TPUs and TensorFlow are fast because they flow through matrix multiplications directly without storing intermediate values in memory.

2.5.2 *Statistical physics*

A convenient theoretical approach to SNFT is based on statistical physics. Statistical physics is selected because of its focus on probability and inference in large systems. The two model systems used to elaborate the SNFT are also based on statistical models in physical systems (the brain and disordered magnets).

> *All of physics in my view, will be seen someday to follow the pattern of thermodynamics and statistical mechanics*
>
> — John Archibald Wheeler (1983, p. 398)

Statistical physics is a catchall that includes statistical mechanics, probability, and thermodynamics. The great benefit of statistical physics is that it provides a generalized method, based in probability, for linking microscopic noise to macroscopic labels (Mayants, 1984, p. 174). Smart networks are also fundamentally based on probability. Smart network technologies such as blockchain and deep learning are probabilistic state machines that coordinate thousands to millions of constituent elements (processing nodes, whether perceptrons or miners, which can be seen as particles) to make high-probability guesses about reality states of the world. Hence, statistical physics could be a good basis for the formalization of smart networks.

Maxwell was among the first to suggest the application of probability as a general model for the study of the science of the very small. Statistical mechanics was likewise intended as a general method based on probability in its early development by Gibbs, following on from classical mechanics (Gibbs, 1902). The aim of statistical mechanics is to address all aspects of mechanical systems, at both the microscopic and macroscopic levels, for example, to explain transitions between gaseous and non-gaseous states.

The thermodynamic aspect of statistical physics is relevant in the formulation of SNFTs because smart networks are physical systems with thermodynamic effects. Blockchains are worldwide physical network systems, comprising about 10,000 nodes on average hosting the transaction ledger for Bitcoin [10,403 (Bitnodes, 2019)] and Ethereum [8,141 (Ethernodes, 2019)]. Deep learning networks too have a physical basis, in that they run on dedicated hardware systems (NVIDIA GPU networks and Google TPU clusters). Concepts such as work, heat, and energy have thermodynamical measures in smart network systems. Blockchains perform work in the form of consensus algorithms (proof-of-work, proof-of-stake, etc.) being a primary mechanism for providing network security and updating the ledger balances of the distributed computing system. Deep learning networks also perform work in the sense of running an operating cycle to derive a predictive classification model for data. The network expounds significant resources to iteratively cycle forward and back through the layers to optimize trial-and-error guesses about the weighting of relevant abstracted feature sets such that new data can be correctly identified.

Technophysics formulations of blockchains and deep learning have been proposed on the basis of thermodynamic properties. For example, a blockchain proof-of-work consensus process could be instantiated as an energy optimization problem with Hamiltonian optimizers and executed as a quantum annealing process on quantum computers (Kalinin & Berloff, 2018). In deep learning, a thermodynamics of machine learning approach is used to propose representation learning as an alternative framework for reasoning in machine learning systems, whose distortion could be measured as a thermodynamical quantity (Alemi & Fischer, 2018).

2.6 Field Theory

A field theory is a theory that describes a background space and how the constituent elements in the space behave. In classical physics, a field is a region in which each point is affected by a physical quantity, be it a force, a temperature or any other scalar, vector, or tensor quantity. For example, objects fall to the ground because they are affected by the force of Earth's gravitational field. A field is a region of space that is affected by a physical quantity that can be represented with a number or a tensor (multi-dimensional number), that has a value for each point in space and time. A weather map, for example, has a temperature assigned to each point on a map. The temperatures may be studied at a fixed point in time (today's temperature) or over a time interval to understand the dynamics of the system (the effects of temperature change).

Field theories are a particularly good mechanism for studying the dynamics of a system. The dynamics of a system refers to how a system changes with time or with respect to other independent physical variables upon which the system depends. The dynamics are obtained by writing an equation called a Lagrangian or a Hamiltonian of the field, and treating it as a classical or quantum mechanical system, possibly with an infinite number of degrees of freedom (parameters). The resulting field theories are referred to as classical or quantum field theories. The dynamics of a classical field are typically specified by the Lagrangian density in terms of the field components; the dynamics can be obtained by using the action principle. The dynamics of a quantum field are more complicated. However, since quantum mechanics may underlie all physical phenomena, it should be possible to cast a classical field theory in quantum mechanical terms, at least in principle, and this is assumed in the SNFT and SNQFT constructions.

2.6.1 *The field is the fundamental building block of reality*

At the quantum mechanical scale, the intuition behind field theory is that fields, not particles, may be the fundamental building blocks of reality. For example, Feynman points out that in the modern framework of the quantum theory of fields, even without referring to a test particle, a field occupies space, contains energy, and its presence precludes a classical true vacuum. This has led physicists to consider fields to be physical entities and a foundational aspect of quantum mechanical systems. The fact that the electromagnetic field can possess momentum and energy makes it very real (Feynman, 1970). One resulting interpretation is that fields underlie particles. Particles are produced as waves or excitations of so-called matter fields. Reality may be composed of fluid-like substances (having properties of flow) called fields. Quantum mechanical reality may be made up of fields, not particles.

2.6.2 *Field theories: Fundamental or effective*

Theories are either fundamental or effective. Fundamental theories are foundational universal truths and effective theories are reasonably effective theories, given the absence of additional proof or knowledge. Fundamental theories have the tony weight of absolute truth. Effective theories serve effectively in the sense of being a reasonable approximation of situations that are not yet fully understood. Classical theories of physics were initially thought to be fundamental, but then found not to be valid everywhere in the universe. Newtonian physics describes pulleys, but not electrons or detailed planetary movement (Einsteinian physics is used in GPS technology). In this sense, all theories of nature are effective theories, in that each is a possible approximation of some more fundamental theory that is as yet unknown.

There is another sense of the meaning of effective field theories, which is that the theory is only effective within a certain range. An effective theory may only be true within certain parameters or regimes, typically whatever domain or length-scale is used to experimentally verify the theory (Williams, 2017). For example, an effective field theory is a way to describe what happens at low energies and long wavelengths (in the domain of general relativity) without having a complete picture of what is happening at higher energies (in the domain of quantum mechanics). In high-energy physics (particle physics), processes can be calculated with the so-called Standard Model without needing to have a complete picture of grand unification or quantum gravity. The opposite is also true, when calculating problems in low-energy physics (gravitational waves), the effects of higher-energy physics (particle physics) can be bracketed out or summed up with a few measurable parameters (Carroll *et al.*, 2014). Each domain has field theories that are effective within its scale-range. The difficulty is deriving field theories that explain situations in which high-energy physics and low-energy physics come together such as black holes.

2.6.2.1 *Effective field theories in quantum mechanics*

Whereas a classical field theory is a theory of classical fields, a quantum field theory is a theory of quantum mechanical fields. A classical field theory is typically specified in conventional space and time (the 3D space and time of Euclidean macroscale reality). On the other hand, a quantum field theory is specified on some kind of background of different models of space and time. To reduce complexity, quantum field theories are most generically placed on either a fixed background such as a flat space, or a Minkowski space (3D quantum mechanical space–time). Whatever space and time region in which the quantum field theory is specified, the idea is to quantize the geometry and the matter contents of the quantum field into an effective theory that can be used to perform calculations. Effective field theories are useful because they can span classical and quantum domains, and more generally, different levels in systems with phase transitions. The SNFT is both a classical field theory and a quantum field theory.

2.6.3 *The smart network theories are effective field theories*

The SNFTs start with the idea that an effective field theory is a type of approximation, or effective theory, for an underlying physical theory (smart networks in this case). The effective field theory is a precision tool that can be used to isolate and explain a relevant part of a system in simpler terms that are analytically solvable. An effective field theory includes the appropriate degrees of freedom (parameters) to describe the physical phenomena occurring at a chosen length-scale or energy-scale within a system, while ignoring substructure and degrees of freedom at other distances or energies (Giorgi *et al.*, 2004). The strategy is to average over the behavior of the underlying theory at shorter length-scales to derive what is hoped to be a simplified model for longer length-scales, which applies to the overall system.

Effective field theories connote the existence of different scale levels within a system. They have been used to explain domains and simplify problems in many areas of particle physics, statistical mechanics, condensed matter physics, superconductivity, general relativity, and hydrodynamics. In condensed matter physics, effective field theories can be used, for example, to study multi-electron atoms, for which solving the Schrödinger equation is not feasible. In particle physics, effective field theories attempt to explain problems such as the Fermi theory of beta decay. In general relativity, effective field theories have been used to simplify gravitational wave problems, and theorize that general relativity itself may be the low-energy effective field theory of a full theory of quantum gravity (in which the expansion scale is the Planck mass). Particularly relevant for quantum computing is the practical application of effective field theories in the domains of superconductivity and condensed matter physics.

2.6.4 *Complex multi-level systems*

A key requirement for a SNFT is that it can be used to manage across diverse scale levels within a complex system. Such a field theory should be able to "identify macroscopic smoothness from microscopic noise" as prescribed by complexity theory (Mitchell, 2009). Various methods, including statistical physics, may be used for linking multiple dimensions within complex systems to obtain signal from noise.

Some aspects of a system are easier to measure on different scales. For example, computing the energy spectrum of the Hamiltonian at different levels of quantum mechanical systems can be challenging. Such calculations may be straightforward at higher levels of the system abstraction, but more difficult when incorporating the energetic fields in which the particles actually behave. At this scale, it is essentially impossible to compute because there is so much data about particle movement. One strategy is to reinterpret particles as states of a quantized field (Jaffe & Witten, 2001). A field theory helps to reinstantiate or roll the system up to a higher level of abstraction at which such calculations can be made.

The method is finding or defining an effective field theory at a scale that renders the system analytically solvable.

For example, the elliptical orbits of the planets are more easily calculated with Newtonian gravity than with general relativity. This simplification can be all that is necessary for certain applications. The benefit of a field theory is that is provides the ability to focus on a particular scale of a system, emphasizing one aspect while limiting others (Georgi, 1993). The objective is to find the simplest framework that captures the essential physics of the target area. For example, when there is interest in lighter particles (such as bottom quarks), heavier particles (e.g. z-bosons and w-bosons) can be eliminated from the model.

In complex multi-level systems, identifying a macroscopic term corresponding to microscopic behavior is a key challenge. The analogs to the temperature and pressure terms arising from a room of septillions of moving particles in a model system are not always clear. Hence, an effective field theory is a formal process that can be used to identify a system's "temperature" term and other system-level metrics. Effective field theories are similar to the renormalization concept (in the sense of mathematically scaling to a different level of the system to focus on a parameter of interest in a simplified manner that can be calculated).

2.7 Five Steps to Defining an Effective Field Theory

Effective field theories are important because there is interesting physics at all scales. Being able to portably travel up and down scale dimensions can make it easier to analyze certain aspects of systems. The idea is to use effective field theories as a tool for isolating parameters of interest within a system and engaging the system at that level. Effective field theories may work best when there is a large separation between the length scale of interest and the length scale of the underlying dynamics.

A distillation of the steps involved in deriving an effective field theory is outlined in Table 2.5. The aim of an effective field theory is to specify the simplest framework that captures the essential physics of interest. The zeroth step is to confirm that there are no already existing fundamental theories to describe the phenomenon and that an effective

Table 2.5. Steps in articulating an effective field theory.

Step	Description
1. Define the system	Characterize the overall scope, shape, and levels of the system, including the relevant scales, lengths, and energies.
2. Identify system elements	Identify the constituent elements of the system, the kinds of interactions between them.
3. Isolate variables of interest	Articulate the aspects of interest that the field theory should study.
4. Reduce complexity by eliminating unnecessary system substructure	Identity the degrees of freedom (the aspects of the system that matter for the problem of study), and irrelevant substructure that can be ignored, and note symmetries, anomalies, or other known complexity attributes.
5. Identify quantitative metrics	Articulate the mathematics to measure the system, averaging the underlying behavior to derive a simplified model with a global term such as a Hamiltonian or Lagrangian.

Source: Adapted from Manohar (2017).

field theory is useful. Theories with related aspects could be identified as inspiration.

The first of the five steps is to define the system by characterizing the overall scope and shape of the system to be studied, including the relevant scale levels in terms of lengths or energies that comprise the system. The second step is to identify the system elements, the particles or other elements that constitute the system, and the kinds of interactions between them. The third step is to isolate the particular variables of interest that the theory aims to study. The fourth step is to reduce complexity by eliminating the unnecessary system substructure which can be ignored for studying the variables of interest within the system. More detailed aspects of the subsystem of interest are identified such as the degrees of freedom (system parameters) and any complexity properties such as symmetry and anomalies that may influence the theory application. The fifth step is identifying the relevant quantitative metrics for measuring the system. The available quantities in the system are identified, and averaged over to generate a metric as a system composite measure such as a Hamiltonian or Lagrangian.

An effective field theory example in a biological neural network is that the system-wide quantity of interest might be the spiking activation (the threshold at which neurons fire), and other data would be superfluous. Another example is a minimal effective field theory that only specifies the fields, the interactions, and the power counting of the system (the dimensions of power counting across scales).

Beyond the basic steps, effective field theories might include more complicated aspects. There could be additional quantities to measure such as available potential energy, propagation, and the range of system states. Also relevant is identifying the dynamics of the system, the dimensions into which the system is expanding. There could be various degrees of freedom. The term degrees of freedom generally connotes system parameters. More specifically, degrees of freedom is a statistical term that means each of a number of independently variable factors that can affect the range of states in which a system may exist, and the directions in which independent motion can occur. Degrees of freedom can be conceived simply as system parameters, and with greater sophistication as a statistical measure of the number of states and ways in which a dynamic system can move, whether this "motion" is considered in physical space or in an abstract space of configurations.

Overall, the key steps in specifying an effective field theory consist of defining (1) the system, (2) the system elements and interactions, (3) the variables of interest, (4) the irrelevant structure that can be ignored, and (5) the quantitative metrics that can be averaged over the system to produce a temperature-type term. Applying the effective field theory development technique to the smart network context, the idea is to consider the different levels and dimensions of the system, and identify the elements, interactions, and relevant quantities to calculate in order to obtain the system behavior. This is developed in more detail in Chapters 11 and 12.

References

Aldridge, I. & Krawciw, S. (2017). *Real-Time Risk: What Investors Should Know About Fintech, High-Frequency Trading and Flash Crashes*. Hoboken, NJ: Wiley.

Alemi, A.A. & Fischer, I. (2018). TherML: Thermodynamics of Machine Learning. ICML 2018. Theoretical Foundations and Applications of Deep Generative Models Workshop.

Bitnodes. https://bitnodes.earn.com/. Accessed June 30, 2019.

Boyden, E.S. (2015). Optogenetics and the future of neuroscience. *Nat. Neurosci.* 18:1200–1.

Brin, D. (2002). *Kiln People*. New York, NY: Tor Books.

Carroll, S.M., Leichenauer, S. & Pollack, J. (2014). A consistent effective theory of long-wavelength cosmological perturbations. *Phys. Rev. D.* 90:023518.

Chen, H. & Ho, K. (2018). Integrated space logistics mission planning and spacecraft design with mixed-integer nonlinear programming. *J. Spacecr. Rockets.* 55(2):365–81.

Dorfler, F., Chertkov, M. & Bullo, F. (2013). Synchronization in complex oscillator networks and smart grids. *PNAS* 110(6):2005–10.

Duong, T., Todi, K.K. & Chaudhary, U. (2019). Decentralizing air traffic flow management with blockchain-based reinforcement learning. Aalto University, Finland.

Ethernodes. https://ethernodes.org/network/1. Accessed June 30, 2019.

Ferrer, E.C. (2017). The blockchain: A new framework for robotic swarm systems. arXiv:1608.00695 [cs.RO].

Ferrer, E.C., Hardjono, T., Dorigo, M. & Pentland, A. (2019). Secure and secret cooperation of robotic swarms by using Merkle trees. arXiv:1904.09266 [cs.RO].

Feynman, R.P. (1970). *The Feynman Lectures on Physics*. Vol I. London, UK: Pearson PTR.

Georgi, H. (1993). Effective field theory. *Annu. Rev. Nucl. Part. Sci.* 43:209–52.

Gibbs, J.W. (1902). *Elementary Principles in Statistical Mechanics*. New York, NY: Scribner.

Giorgi, G.A, Guerraggio A. & Thierfelder J. (2004). *Mathematics of Optimization: Smooth and Nonsmooth Case*. London, UK: Elsevier.

Hammi, B., Khatoun, R., Zeadally, S. *et al.* (2017). Internet of Things (IoT) technologies for smart cities. *IET Networks*. 7.

Jaffe, A & Witten, E. (2001). Quantum Yang–Mills Theory, 1–14. https://www.claymath.org/sites/default/files/yangmills.pdf. Accessed June 30, 2019.

Kalinin, K.P. & Berloff, N.G. (2018). Blockchain platform with proof-of-work based on analog Hamiltonian optimisers. arXiv:1802.10091 [quant-ph].

Kokar, M.M., Baclawski, K. & Eracar, Y.A. (1999). Control theory-based foundations of self-controlling software. *IEEE Intell. Syst. App.* 14(3):37–45.

Manohar, A.V. (2017). Introduction to effective field theories. *EFT (Particle Physics and Cosmology)* July 3–28, 1–94.

Martins, N.R.B., Angelica, A., Chakravarthy, K. *et al.* (2019). Human brain/cloud interface. *Front. Neurosci.* 13:112.

Mayants, L. (1984). *The Enigma of Probability and Physics.* New York, NY: Springer.

Mitchell, M. (2009). *Complexity: A Guided Tour.* Oxford, UK: Oxford University Press.

Pieroni, A., Scarpato, N., Di Nunzio, L. *et al.* (2018). Smarter city: smart energy grid based on blockchain technology. *Intl. J. Adv. Sci. Eng. Info. Tech.* 8(1):298–306.

SAE (2018). SAE International Releases Updated Visual Chart for Its "Levels of Driving Automation" Standard for Self-Driving Vehicles. SAE.

Sayedi, A. (2018). Real-time bidding in online display advertising. *Market. Sci.* 37(4):553–68.

Singh, S., Lu, S., Kokar, M.M. & Kogut, P.A. (2017). Detection and classification of emergent behaviors using multi-agent simulation framework (WIP). Spring 2017 Simulation Multi-Conference (SCS).

Swan, M. (2015). *Blockchain: Blueprint for a New Economy.* Sebastopol, CA: O'Reilly Media.

Swan, M. (2016). The future of brain-computer interfaces: Blockchaining your way into a cloudmind. *JET.* 26(2).

Swan, M. (2018). Blockchain for business: Next-generation enterprise artificial intelligence systems. In: Raj, P., Deka, G.C. (eds). *Advances in Computers,* Vol. 111. *Blockchain Technology: Platforms, Tools and Use Cases.* London, UK: Elsevier.

Swan, M. (forthcoming). Technophysics, Smart health networks, and the bio-cryptoeconomy. In: Boehm, F. (ed). *Nanotechnology, Nanomedicine, and AI: Toward the Dream of Global Health Care Equivalency.* Boca Raton, FL: CRC Press.

Tasca, P. & Tessone, C.J. (2019). A taxonomy of blockchain technologies: Principles of identification and classification. *Ledger* 4.

Vasarhelyi, G., Viragh, C., Somorjai, G. *et al.* (2018). Optimized flocking of autonomous drones in confined environments. *Sci. Robot.* 3(20):eaat3536.

Wheeler, J.A. (1983). On recognizing 'law without law.' Oersted Medal Response at the joint APS-AAPT Meeting, New York, 25 January 1983. *Am. J. Phys.* 51(5):398–406.

Williams, M. (2017). *Effective Theory,* Motifs in Physics Series. Harvard University, pp. 1–6.

Woerner, K., Benjamin, M.R., Novitzky, M. & Leonard, J.J. (2019). Quantifying protocol evaluation for autonomous collision avoidance. *Auton. Robots.* 43(4):967–91.

Chapter 3

Quantum Computing: Basic Concepts

... it seems that the laws of physics present no barrier to reducing the size of computers until bits are the size of atoms, and quantum behavior holds sway

— Richard P. Feynman (1985)

Abstract

Quantum computing is a research frontier in physical science with a focus on developing information processing at the quantum scale. Quantum computing involves the use of algorithms to exploit the special properties of quantum mechanical objects (such as superposition, entanglement, and interference) to perform computation. In physics, quantum mechanics is the body of laws that describe the behavior and interaction of electrons, photons, and other subatomic particles that make up the universe. Quantum computing engages the rules of quantum mechanics to solve problems using quantum information. Quantum information is information concerning the state of a quantum system which can be manipulated using quantum information algorithms and other processing techniques. Although quantum computing is farther along than may be widely known, it is an early-stage technology fraught with uncertainty. The overall aim in the longer term is to construct universal fault-tolerant quantum computers.

3.1 Introduction

Quantum computers are in the early stages of development and would likely be complementary to existing computational infrastructure, interacting with classical devices, and being accessed either locally or as a cloud service. Currently, the top methods demonstrate 30–70 qubits of processing power and achieve fidelity rates above 99% (i.e. below a fault tolerance threshold of 1%). However, there is uncertainty about the realizability of scalable universal quantum computers. Quantum computers may excel at solving certain types of problems such as optimization. This could offer a step-up in computing such that it is possible to solve new classes of problems, but not all problems. For example, considering well-known optimization problems, it may be possible to search twice as many possibilities in half the time (exploring a fixed-size space in the square root of the amount of time required for a classical computer).

Quantum computing is an early-stage technology with numerous risks and limitations (Dyakonov, 2018). The long-term goal of universal quantum computing is not immediate as many challenges including error correction need to be resolved. In the short term, the focus is on solving simple problems in which quantum computers offer an advantage over classical methods through NISQ devices (noisy intermediate-scale quantum devices) (Preskill, 2018).

3.1.1 *Breaking RSA encryption*

One of the biggest questions is when it might be possible to break existing cryptography standards with quantum computing. The current standard is 2048-bit RSA (Rivest–Shamir–Adleman) encryption, which is widely used for activities such as securely sending credit card details over the internet. Predictions vary as to when it may be possible to break the current standard, meaning factor the 2048-bit integers used by the RSA method. Although not immanent, methods are constantly improving, and readying cryptographic systems for the quantum era is prescribed.

A 2019 report published by the US National Academies of Sciences predicts that breaking RSA encryption is unlikely within the next decade. The report indicates that any serious applications of quantum computing are at least 10 years away (Grumbling & Horowitz, 2019). Given the current state of quantum computing and the recent rates of progress, "it is highly unexpected that a quantum computer that can compromise RSA 2048 or comparable discrete logarithm-based public key cryptosystems will be built within the next decade" (Grumbling & Horowitz, 2019, 157). The report's stance on error correction is that "The average error rate of qubits in today's larger devices would need to be reduced by a factor of 10 to 100 before a computation could be robust enough to support [the required] error correction at scale" (Grumbling & Horowitz, 2019). The report further highlights that "at this error rate, the number of physical qubits held by these devices would need to increase at least by a factor of 10^5 in order to create a useful number of effective logical qubits" (Grumbling & Horowitz, 2019). A 2016 National Institute of Standards and Technology (NIST) report has a similar result, noting that some of the most aggressive time estimates predicting when quantum computers might be powerful enough to break 2048-bit RSA might be by 2030, at a potential cost of a billion dollars (Chen *et al.*, 2016, 6).

One complication in making predictions is that the number of required qubits (processing power) needed to factor 2048-bit RSA integers varies by method. Different algorithms need different numbers of qubits (Mavroeidis *et al.*, 2018). Although difficult to guess, "current estimates range from tens of millions to a billion physical qubits" (Mosca, 2018, 39). Newer estimates propose more granularity, indicating in more detail how a quantum computer might perform the calculation with 20 million noisy qubits (without error correction) in just 8 hours (Gidney & Ekera, 2019). (The method relies on modular exponentiation which is the most computationally expensive operation in Shor's algorithm for factoring.) In 2012, a 4-qubit quantum computer factored the number 143, and in 2014, a similar device factored the number 56,153. However, scaling up quickly is not straightforward because the extent of error correction required is unknown. The recent result from Gidney & Ekera (2019) suggesting that 20 million qubits might be possible without error correction is potentially a landmark step.

One of the nearest term remedies for post-quantum security is quantum cryptography, in the form of quantum key distribution (QKD), which has been foreseen in quantum information science roadmaps for some time (Los Alamos National Laboratory (LANL), 2004). QKD is the idea of issuing cryptographic keys generated with quantum computing methods and distributing them via global communications networks, satellite-based and terrestrial. QKD has been experimentally demonstrated and remains to be commercialized. The market for QKD is estimated to reach $980 million by 2024 from $85 million in 2019 (Gasman, 2019).

3.2 Basic Concepts: Bit and Qubit

Quantum computing springs from Nobel physicist Richard Feynman's intuition that it should be possible to perform very powerful computations by using quantum building blocks. He suggests the idea of simulating physics with computers using a universal quantum simulator.

Feynman worked on many problems at the small scales of the quantum mechanical domain. He famously announced that "There's plenty of room at the bottom" (Feynman, 1960), referring to the idea of miniaturization at the atomic scale such that the entire 24-volume set of the *Encyclopedia Britannica* might be printed on the head of a pin. These ideas helped to sponsor the nanotechnology industry in the 1990s and are likewise motivating the current development of quantum computing. To get an idea of scale, the nanometer scale is 10^{-9} m, the atomic scale (in terms of the Bohr radius being the probable distance from the nucleus to the electron) is 10^{-11} m, and the electron scale (the size of the electron) is 10^{-15} m.

Feynman suggested that a quantum computer could be an efficient universal simulator of quantum mechanics. Such a "universal quantum simulator" (Feynman, 1982, 474) would be a different kind of computer that is not a traditional Turing machine. He posits two ways to simulate quantum mechanics with computers. One is reconceiving the notion of computers and building computers out of quantum mechanical elements that obey quantum mechanical laws. The other idea is trying to imitate quantum mechanical systems with classical systems.

Feynman's key thought is that the more closely computing systems can be built in the structure of nature, the better they can simulate nature. He says that the "the various field theories have the same kind of behavior, and can be simulated in every way, apparently, with little latticeworks of spins and other things" (Feynman, 1982, 474–5). Assuming that the world is made in a discrete lattice, "the phenomena of field theories can be well imitated by many phenomena in solid state theory (which is simply the analysis of a latticework of crystal atoms, and in the case of the kind of solid state I mean each atom is just a point which has numbers associated with it, with quantum mechanical rules)" (Feynman, 1982, 475).

Feynman proposed the idea of the universal quantum simulator in 1982, following which there have been other theoretical developments. In 1985, David Deutsch proposed a universal quantum computer based on the idea that quantum gates could function in a similar fashion to traditional digital computing binary logic gates [Deutsch, 1985]. In 2000, Charlie Bennett showed that it is possible to efficiently simulate any classical computation using a quantum computer (Bennett & DiVincenzo, 2000).

Other advances in recent decades have led to the practical realizability of quantum computers. First, in the 1990s was the discovery of quantum error correction. Unlike classical bits that persistently stay in a 1 or 0 state, quantum bits are extremely sensitive to environmental noise and may decohere before they can be used to perform a computation. Quantum error correction overcomes some of the challenges of working in quantum mechanical domains.

Second, since 2012, there have been advances in room-temperature superconducting materials and a proliferation of ways of making qubits such that quantum systems have increased from 1–2 qubits to 50–100 qubits. A research goal is demonstrating *quantum advantage*, which is specific cases in which quantum computing confers an advantage over classical computing.

3.2.1 *Quantum computing and classical computing*

Quantum information processing is not only a potentially faster means of computing but also a new paradigm in that information is conceived and managed in a completely different way due to the different properties of quantum objects. According to W.D. Phillips, 1997 Nobel Prize winner in physics and NIST scientist, "Quantum information is a radical departure in information technology, more fundamentally different from current technology than the digital computer is from the abacus" (Williams, 2007).

Some of the special properties of quantum objects (be it atoms, ions, or photons) are superposition, entanglement, and interference (SEI properties). Superposition means that particles can exist across all possible states simultaneously. This is known as a superposition of states. For example, an electron may exist in two possible spin states simultaneously, referred to as 1 and 0, or spin-up and spin-down. Entanglement is the situation that groups of particles are related and can interact in ways such that the quantum state of each particle cannot be described independently of the state of the others even when the particles are separated by a large distance. Across large distances, this is called Bell pair entanglement or nonlocality. Interference relates to the wave-like behavior of particles. Interference can be positive or negative, in that when two waves come together, they are either reinforced or diminished.

Classical computing is based on electrical conductivity, using Boolean algebra (namely expressions evaluating as true/false, and/or, etc.) to manipulate bits. Quantum computing is based on quantum mechanics, using vectors and linear algebra to manipulate matrices of complex numbers. Aiming toward a universal model of quantum computation, the idea is to package the quantum mechanical matrix manipulations such that they run quantum states that are executed with a set of gates that offer the same kind of Boolean logic as in classical computing.

3.2.2 *Bit and qubit*

In classical computing, the bit is the fundamental computational unit. The bit is an abstract mathematical entity that is either a 0 or a 1. Computations are constructed as a series of manipulations of 0s and 1s. In the physical world, a bit might be represented in terms of a voltage inside a computer, a magnetic domain on a hard disk, or light in an optical fiber. The qubit (quantum bit) is the equivalent system in quantum mechanics. The qubit is likewise an abstract mathematical entity (a logical qubit), existing in a superposition state of being both a 0 and a 1, until collapsed in the measurement at the end of the computation into being a classical 0 or 1. The qubit can be instantiated in different ways in the physical world. There are realizations of qubits in atoms, photons, electrons, and other kinds of physical systems. The quantum state of a qubit is a vector in a 2D space.

This is a linear combination of the 1 and the 0 (the trajectory or probability that it is in the 1 or the 0 state). A model of computation can be built up by assigning states closer to the 0 as being 0 and states closer to the 1 as being 1 (when measured).

A bit is always in a state of either 1 or 0. A qubit exists in a state of being both 1 and 0 until it is collapsed into a 1 or a 0 at the end of the computation. A bit is a classical object that exists in an electronic circuit register. A qubit is a quantum object (an atom, photon, or electron) that bounces around in a 3D space with a different probability of being at any particular place in the 3D sphere called a Hilbert space (and has vector coordinates in the X, Y, and Z directions). Figure 3.1 shows the physical space of the states of the bit and the qubit.

The interpretation is that whereas a classical bit is either on or off (in the state of 1 or 0), a qubit can be on and off (1 and 0) at the same time, a property called superposition. One example of this is the spin of the electron in which the two levels can be understood as spin-up and spin-down. Another example is the polarization of a single photon in which the two states can be taken to be the vertical polarization and the horizontal polarization (single photons are often transmitted in communications networks on the basis of polarization). In a classical system, a bit needs to be in one state or the other. However, in a quantum mechanical system, the qubit can be in a coherent superposition of both states or levels of the system simultaneously, a property which is fundamental to quantum mechanics and indicates the greater potential range of computation in quantum systems.

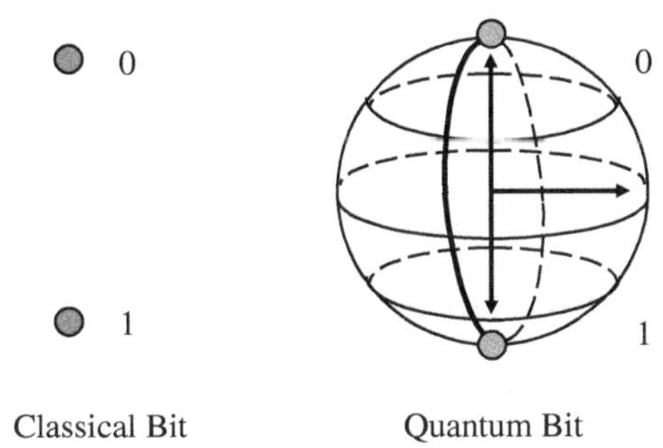

Figure 3.1. Potential states of bit and qubit.

Compared to classical states, quantum states are much richer and have more depth. Superposition means that quantum states can have weight in all possible classical states. Each step in the execution of a quantum algorithm mixes the states into more complex superpositions. For example, starting with the qubit in a position of 0–0–0 leads to a superposition of 1–0–0, 1–0–1, and 1–1–1. Then each of the three parts of the superposition state branches out into even more states. This indicates the extensibility of quantum computers that could allow faster problem solving than is available in classical computers.

3.2.3 *Creating qubits*

A qubit can be created in any quantum system which has two levels of energy that can be manipulated (Steane, 1997). Qubits can be conceived as being similar to harmonic oscillators at the macroscale. Physical systems that vibrate in a wave-like form between two levels of energy are called harmonic oscillators. Some examples include electrical circuits with oscillating current, sound waves in gas, and pendulums. Harmonic oscillators can be modeled as a wave function that cycles between the peak and trough energy levels. The same wave function concept is true at the quantum scale. In this sense, whenever there is a quantum system with two levels of energy, it can be said to be a qubit and possibly engaged as a two-state quantum device. This implies that there can be many different ways of building qubits. Hence, the method for creating qubits might be an engineering choice similar to the way that different methods have been used in classical computing for the physical implementation of logic gates (methods have ranged over time and included vacuum tubes, relays, and most recently integrated circuits).

3.3 Quantum Hardware Approaches

3.3.1 *The DiVincenzo criteria*

The DiVincenzo criteria have been proposed as standards that constitute the five elements of producing a well-formed quantum computer (DiVincenzo, 2000). The criteria are having (1) a scalable system of well-characterized qubits, (2) qubits that can be initialized with fidelity (typically to the zero state), (3) qubits that have a long-enough coherence time for the calculation (with low error rates), (4) a universal set of quantum gates (that can be implemented in any system), and (5) the capability of measuring any specific qubit in the ending result.

There are several approaches to quantum computing (Table 3.1) (McMahon, 2018). Those with the most near-term focus are superconducting circuits, ion trapping, topological matter, and quantum photonics. Irrespective of the method, the objective is to produce quantum computing chips that perform computations with qubits, using a series of quantum logic gates that are built into quantum circuits, whose operation is programmed with quantum algorithms. Quantum systems may be accessed locally or as a cloud service. As of June 2019, one method is commercially available, which is superconducting circuits. Verification of computational claims is a considerable concern. External parties such as academic

Table 3.1. Quantum computing hardware platforms.

Organization	Qubit type	# qubits	Commercial status
1. IBM	Superconducting (gate model)	19(50)	Available
2. D-Wave	Superconducting (quantum annealing)	2048	Available
3. Rigetti	Superconducting (gate model)	19	Available
4. Google	Superconducting (gate model)	72	Built, unreleased
5. Intel/Delft	Superconducting	49	Built, unreleased
6. QCI	Superconducting	Unknown	Research
7. IonQ	Trapped ions	23	Built, unreleased
8. Alpine Quantum Tech.	Trapped ions	Unknown	Research
9. Microsoft	Majorana fermion	Unknown	Research
10. Nokia Bell Labs	FQH State	Unknown	Research
11. Xanadu	Photonic	Unknown	Research
12. PsiQuantum	Photonic	Unknown	Research
13. ColdQuanta	Neutral atoms	50	Research
14. HRL	Quantum dots	Unknown	Research
15. SQC	Quantum dots	Unknown	Research
16. NMR Cloud Q	NMR	4	Research

scientists are engaged to confirm, verify, and benchmark the results of different quantum systems, for example, for Google (Villalonga *et al.*, 2019) and for IonQ (Murali *et al.*, 2019).

3.3.2 *Superconducting circuits: Standard gate model*

The most prominent approach to quantum computing is superconducting circuits. Qubits are formed by an electrical circuit with oscillating current and controlled by electromagnetic fields. Superconductors are materials which have zero electrical resistance when cooled below a certain temperature. (In fact, it is estimated that more than half of the basic elements in the periodic table become superconducting if they are cooled to sufficiently low temperatures.) Mastering superconducting materials could be quite useful since as a general rule, about 20% of electricity is lost due to resistance. The benefit of zero electrical resistance for quantum computing is that electrons can travel completely unimpeded without any energy dissipation. When the temperature drops below the critical level, two electrons (which usually repel each another) form a weak bond and become a so-called Cooper pair that experiences no resistance when going through metal (tunneling) and which can be manipulated in quantum computing.

Superconducting materials are used in quantum computing to produce superconducting circuits that look architecturally similar to classical computing circuits, but are made from qubits. There is an electrical circuit with oscillating current in the shape of a superconducting loop that has the circulating current and a corresponding magnetic field that can hold the qubits in place. Current is passed through the superconducting loop in both directions to create the two states of the qubit. More technically, the superconducting loop is a superconducting quantum interference device (SQUID) magnetometer (a device for measuring magnetic fields), which has two superconductors separated by thin insulating layers to form two parallel Josephson junctions. Josephson junctions are key to quantum computing because they are nonlinear superconducting inductors that create the energy levels needed to make a distinct qubit.

Specifically, the nonlinearity of the Josephson inductance breaks the degeneracy of the energy-level spacings, allowing the dynamics of the system to be restricted to only the 2-qubit states. The Josephson junctions are necessary to produce the qubits; otherwise, the superconducting loop would just be a circuit. The point is that the linear inductors in a traditional circuit are replaced with the Josephson junction, which is a nonlinear element that produces energy levels with different spacings from each other that can be used as a qubit. Josephson (after whom the Josephson junction is named) was awarded the Nobel Prize in Physics in 1973 for work predicting the tunneling behavior of superconducting Cooper pairs.

As an example of a superconducting system, Google's qubits are electrical oscillators constructed from aluminum (niobium is also used), which becomes superconducting when cooled to below 1 K (−272°C). The oscillators store small amounts of electrical energy. When the oscillator is in the 0 state, it has zero energy, and when the oscillator is in the 1 state, it has a single quantum of energy. The two states of the oscillator with 0 or 1 quantum of energy are the logical states of the qubit. The resonance frequency of the oscillators is 6 gigahertz (which corresponds to 300 millikelvin) and sets the energy differential between the 0 and 1 states. The frequency is low enough so that control electronics can be built from readily available commercial components and also high enough so that the ambient thermal energy does not scramble the oscillation and introduce errors. In another example, Rigetti has a different architecture. This system consists of a single Josephson junction qubit on a sapphire substrate. The substrate is embedded in a copper waveguide cavity. The waveguide is coupled to qubit transitions to perform quantum computations (Rigetti *et al.*, 2012).

3.3.2.1 *Superconducting materials*

Superconducting materials is an active area of ongoing research (Table 3.2). The discovery of "high-temperature superconductors" in 1986 led to the feasibility of using superconducting circuits in quantum computing (and the 1987 Nobel Prize in Physics) (Bednorz & Muller, 1986). Before high-temperature superconductors, ordinary superconductors were known materials that become superconducting at critical temperatures below 30 K (−303°C), when cooled with liquid helium. High-temperature superconductors constitute advanced materials because transition temperatures can be

Table 3.2. Superconducting materials.

Superconducting material	Critical temperature (K)	Critical temperature (°C)	Discovery
Ordinary superconducting materials	Below 30	−303	1911
High-temperature superconducting materials	138	−135	1986
Room-temperature superconducting materials	203	−70	2015
High room-temperature superconducting materials	260	−13	2019

as high as 138 K (−135°C), and materials can be cooled to superconductivity with liquid nitrogen instead of helium. Initially, only certain compounds of copper and oxygen were found to have high-temperature superconducting properties (for example, varieties of copper oxide compounds such as bismuth strontium calcium copper oxide and yttrium barium copper oxide). However, since 2008, several metal-based compounds (such as iron, aluminum, copper, and niobium) have been found to be superconducting at high temperatures too.

Experimental, of interest is a new class of hydrogen-based "room-temperature superconductors" (i.e. warmer than ever before) that have been discovered with high-pressure techniques. In 2015, hydrogen sulfide subjected to extremely high pressure (about 150 gigapascals) was found to have a superconducting transition near 203 K (−70°C) (Drozdov *et al.*, 2015). In 2019, another project produced evidence for superconductivity above 260 K (−13°C) in lanthanum superhydride at megabar pressures [Somayazulu *et al.*, 2019]. Although experimentally demonstrated, such methods are far from development into practical use due to the specialized conditions required to generate them (a small amount of material is pressed between two high-pressure diamond points (Zurek, 2019)).

3.3.3 *Superconducting circuits: Quantum annealing machines*

Within the superconducting circuits approach to quantum computing, there are two architectures, the standard gate model (described above) and quantum annealing (invented first, but more limited). The two models are used for solving different kinds of problems. The universal gate model connotes a general-purpose computer, whereas the annealing machine is specialized. Quantum annealing machines have superconducting qubits with programmable couplings that are designed to solve QUBO problems (quadratic unconstrained binary optimization), a known class of NP-hard optimization problems that minimize a quadratic polynomial over binary variables.

In quantum annealing, the aim is to harness the natural evolution of quantum states over time. A problem is set up at the beginning and then the system runs such that quantum physics takes its natural evolutionary course. There is no control during the system's evolution, and ideally, the ending configuration corresponds to a useful answer to the problem. As compared with quantum annealing, the gate model aims to more fully control and manipulate the evolution of quantum states during the operation. This is more difficult given the sensitivity of quantum mechanical systems, but having more control implies that a bigger and more general range of problems can be solved. The difference in approach explains why quantum annealing machines appeared first and have been able to demonstrate 2048 qubits, whereas only 30–70 qubits are currently achieved in the standard gate model.

Quantum annealing is an energy-based model related to the idea of using the quantum fluctuations of spinning atoms to find the lowest energy state of a system (Kadowaki & Nishimori, 1998). Annealing refers to the centuries-old technique used by blacksmiths to forge iron. In the thermal annealing process, the iron becomes uniformly hot enough so that the atoms settle in the lowest energy landscape, which makes the strongest material. Similarly, quantum annealing is based on the idea of finding the lowest energy configuration of a system.

Quantum annealing is deployed as a method for solving optimization problems by using quantum adiabatic evolution to find the ground state of a system (adiabatic means heat does not enter or leave the system). Run on a quantum computer, the quantum annealing process starts from a ground state which is the quantum mechanical superposition of all possible system states with equal weights. The system then evolves per the time-dependent Schrödinger equation in a natural physical evolution to settle in a low-energy state. The computational problem to be solved is framed in terms of an energy optimization problem in which the low-energy state signals the answer. (The quantum annealing process is described in more detail in Chapter 10.)

Overall, the quantum annealing process allows the system of spins (spinning atoms of qubits) to find a low-energy state. Superconducting circuits in the quantum annealing model can be thought of as programmable annealing engines (Kaminsky & Lloyd, 2004). Optimization problems are framed such that they can be instantiated in the form of an energy landscape minimization. Although annealing machines are not general-purpose quantum computers, one advantage is that since annealing systems constantly attempt to reach the lowest energy state, they are more tolerant and resistant to noise than gate model systems and may require much less error correction at large scales.

3.3.4 *Ion trapping*

Another prominent approach to quantum computing is trapped ions. In these quantum chips, ions are stored in electromagnetic traps and manipulated by lasers and electromagnetic fields. Ions are atoms which have been stripped of or received electrons, which leaves them positively or negatively charged and therefore more easily manipulatable. The advantage of ion trap qubits is that they have a long coherence time (making calculations easier) and (like annealing machines) may require less error correction at large scales. A single qubit trap may accommodate 30–100 qubits, and 23 qubits have been demonstrated in a research context (Murali *et al.*, 2019).

The IonQ quantum chip uses ytterbium ions, which unlike superconducting qubits, do not need to be supercooled to operate. Bulky cryogenic equipment is not required, and the entire system occupies about one cubic meter, as opposed to a much larger footprint for superconducting circuit machines. The chip is a few millimeters across. It is fabricated with silicon and contains 100 electrodes that confine and control the ions in an ultra-high-vacuum environment.

To operate, the ion trap quantum computer holds the ions in a geometrical array (a linear array for IonQ). Laser beams encode and read information to and from individual ions by causing transitions between the electronic states of the ion. The ions influence each other through electrostatic interactions and their coupling can be controlled. More specifically, the IonQ ions form a crystal structure because they repel each other (since they are all of the same isotope of the same element (ytterbium-171)). The electrodes underneath the ions hold the charged particles together in a linear array by applying electrical potentials. The lasers initialize the qubits, entangle them through coupling, and produce quantum logic gates to execute the computation. At the end of the computation, another laser causes ions to fluoresce if they are in a certain qubit state. The fluorescence is collected to measure each qubit and compute the result of the computation. One design principle is already becoming clear in such ion trap systems, that the number of qubits scales as the square root of the gates.

3.3.5 *Majorana fermions and topological quantum computing*

An interesting and somewhat exotic approach for building a universal quantum computer is Majorana fermions. Qubits are made from particles in topological superconductors and electrically controlled in a computational model based on their movement trajectories (called "braiding"). One of the main benefits of topological quantum computing is physical error correction (error correction performed in the hardware, not later by software). The method indicates very low initial error rates as compared with other approaches (Freedman *et al.*, 2002).

Topological superconductors are novel classes of quantum phases that arise in condensed matter, characterized by structures of Cooper pairing states (i.e. quantum computable states) that appear on the topology (the edge and core) of the superconductor (hence the name topological superconductors). The Cooper pairing states are a special class of matter called Majorana fermions (particles identified with their own antiparticles). Topological invariants constrained by the symmetries of the systems produce the Majorana fermions and ensure their stability.

As the Majorana fermions bounce around, their movement trajectories resemble a braid made out of different strands. The braids are wave functions that are used to develop the logic gates in the computation model

(Wang, 2010). Majorana fermions appear in particle–antiparticle pairs and are assigned to quantum states or modes. The computation model is built up around the exchange of the so-called Majorana zero modes in a sequential process. The sequentiality of the process is relevant as changing the order of the exchange operations of the particles changes the final result of the computation. This feature is called non-Abelian, denoting that the steps in the process are non-commuting (non-exchangeable with one another). Majorana zero modes obey a new class of quantum statistics, called non-Abelian statistics, in which the exchange operations of particles are non-commutative.

The Majorana zero modes (modes indicate a specific state of a quantum object related to spin, charge, polarization, or other parameter) are an important and unique state of the Majorana fermionic system (unlike other known bosonic and fermionic matter phases). The benefit of the non-Abelian quantum statistics of the Majorana zero modes is that they can be employed for wave function calculations, namely to average over the particle wave functions in sequential order. The sequential processing of particle wave function behavior is important for constructing efficient logic gates for quantum computation. Researchers indicate that well-separated Majorana zero modes should be able to manifest non-Abelian braiding statistics suitable for unitary gate operations for topological quantum computation (Sarma *et al.*, 2015).

Majorana fermions have only been realized in the specialized conditions of temperatures close to 1 K (–272°C) under high magnetic fields. However, there are recent proposals for more reliable platforms for producing Majorana zero modes (Robinson *et al.*, 2019) and generating more robust Majorana fermions in general (Jack *et al.*, 2019).

3.3.6 *Quantum photonics*

Qubits are formed from either matter (atoms or ions) or light (photons). Quantum photonics is an important approach to quantum computing given its potential linkage to optical networks, in the fact that global communications networks are based on photonic transfer. In quantum photonics, single photons or squeezed states of light in silicon waveguides are used to represent qubits, and they are controlled in a computational model in cluster states (entangled states of multiple photons). Quantum photonics can be realized in computing chips or in free space. Single photons or squeezed states of light are sent through the chip or the free space for the computation and then measured with photon detectors at the other end.

For photonic quantum computing, a cluster state of entangled photons must be produced. The cluster state is a resource state of multi-dimensional highly entangled qubits. There are various ways of generating and using the cluster state (Rudolph, 2016). The general process is to produce photons, entangle them, compute with them, and measure the result. One way of generating cluster states is in lattices of qubits with Ising-type interactions (phase transitions). Lattices translate well into computation. Cluster states are represented as graph states, in which the underlying graph is a connected subset of a d-dimensional lattice. The graph states are then instantiated as a computation graph with directed operations to perform the computation.

3.3.6.1 *Photonic time speed-ups*

The normal speed-up in quantum computing compared to classical computing is due to the superposition of 0s and 1s, in that the quantum circuit can process 0s and 1s at the same time. This provides massive parallelism by being able to process all of the problem inputs at the same time. Photonics allows an additional speed-up to the regular speed-up of quantum computing. In photonic quantum computing, superposition can be used not only for problem inputs but also for processing gates (Procopio *et al.*, 2015). Time can be accelerated by superpositioning the processing gates. Standard quantum architectures have fixed gate arrangements, whereas photonic quantum architectures allow the gate order to be superimposed as well. This means that when computations are executed, they run through circuits that are themselves superpositioned. The potential computational benefit of the superposition of optical quantum circuits is an exponential advantage over classical algorithms and a linear advantage over regular quantum algorithms.

3.3.7 *Neutral atoms, diamond defects, quantum dots, and nuclear magnetic resonance*

Overall, there are many methods for generating qubits and computing with them (Table 3.3). In addition to the four main approaches (superconducting circuits, ion traps, Majorana fermions, and photonics), four additional approaches are discussed briefly. These include neutral atoms, diamond defects (nitrogen-vacancy defect centers), quantum dots, and nuclear magnetic resonance (NMR).

3.3.7.1 *Neutral atoms*

An early-stage approach to quantum computing is neutral atoms. Neutral atoms are regular uncharged atoms with balanced numbers of protons and electrons, as opposed to ions that are charged because they have had

Table 3.3. Qubit types by formation and control parameters.

Qubit type	Qubit formation (DiVincenzo criterion #1)	Qubit control for computation (DiVincenzo criteria #2–5)
1. Superconducting circuits	Electrical circuit with oscillating current	Electromagnetic fields and microwave pulses
2. Trapped ions	Ion (atom stripped of one electron)	Ions stored in electromagnetic traps and manipulated with lasers
3. Majorana fermions	Topological superconductors	Electrically controlled along non-Abelian "braiding" path
4. Photonic circuits	Single photons (or squeezed states) in silicon waveguides	Marshalled cluster state of multi-dimensional entangled qubits
5. Neutral atoms	Electronic states of atoms trapped by laser-formed optical lattice	Controlled by lasers
6. Quantum dots	Electron spins in a semiconductor nanostructure	Microwave pulses
7. Diamond center defects	Defect has an effective spin; the two levels of the spin define a qubit	Microwave fields and lasers

Source: Adapted from McMahon (2018).

an electron stripped away from them or added to them. Qubits are produced by exciting neutral atoms trapped in optical lattices or optical arrays, and qubits are controlled in computation by another set of lasers. The neutral atoms are trapped in space with lasers. An optical lattice is made with interfering laser beams from multiple directions to hold the atoms in wells (an egg carton-shaped structure). Another method is holding the atoms in an array with optical tweezers. Unlike ions (which have strong interactions and repel each other), neutral atoms can be held in close confinement with each other and manipulated in computation. Atoms such as cesium and rubidium are excited into Rydberg states from which they can be manipulated to perform computation (Saffman, 2016). Researchers have been able to accurately program a two-rubidium atom logic gate 97% of the time with the neutral atoms approach (Levine *et al.*, 2018), as compared to 99% fidelity with superconducting qubits. A 3D array of 72 neutral atoms has also been demonstrated (Barredo *et al.*, 2018).

3.3.7.2 *Diamond defects (nitrogen-vacancy defect centers)*

An interesting approach, although one that may have scalability challenges for commercial deployment, is diamond center defects. Imperfections in the crystal lattice within diamonds are commonplace and have been exploited for a variety of uses from crystallography to the development of novel quantum devices. Defects may be the result of natural lattice irregularities or artificially introduced impurities. For quantum computing, impurities are introduced by implanting ions to make nitrogen-vacancy photonic centers. A nitrogen vacancy can be created in a diamond crystal by knocking out a carbon atom and replacing it with a nitrogen atom and also by knocking out a neighboring carbon atom so that there is a vacant spot. The nitrogen vacancy produces the so-called Farbe center (color center), which is a defect in a crystal lattice that is occupied by an unpaired electron. The unpaired electron creates an effective spin which can be manipulated as a qubit. The nitrogen-vacancy defect center is attractive for quantum computing because it produces a robust quantum state that can be initialized, manipulated, and measured with high fidelity at room temperature (Haque & Sumaiya, 2017).

3.3.7.3 *Quantum dots*

Another early-stage approach, in the form of a semiconductor concept, is quantum dots (quantum dots are nanoparticles of semiconducting material) (Loss & DiVincenzo, 1998). In this method, electrically controlled quantum dots that can be used as qubits are created from electron spins trapped in a semiconductor nanostructure, and then electrical pulses are used to control them for computation. A semiconductor-based structure is fabricated that is similar to that of classical processors. Metal electrodes are patterned on the semiconductor layer so that electrostatic fields can be made from the wires to trap single electrons. The spin degrees of freedom of the electrons are used as qubits. Within the semiconductor nanostructure, there are small silicon chambers that keep the electron in place long enough to hybridize its charge and spin and manipulate the electron spin–orbit interactions for computation (Petta *et al.*, 2005). The coherence interactions typically last longer in silicon than in other materials, but can be difficult to control. There has been some improvement in controlling qubit decoherence in quantum dot computing models (Kloeffel & Loss, 2013).

3.3.7.4 *Nuclear magnetic resonance*

Nuclear magnetic resonance (NMR) is one of the first approaches to quantum computing, but is seen as being difficult to scale for commercial purposes. NMR uses the same technology that is used in medical imaging. The physics principle is that since atoms have spin and electrical charge, they may be controlled through the application of an external magnetic field. In 2001, IBM demonstrated the first experimental realization of quantum computing, using NMR (Vandersypen *et al.*, 2001). A 7-qubit circuit performed the simplest instance of Shor's factoring algorithm by factoring the number 15 (into its prime factors of 3 and 5).

References

Barredo, D., Lienhard, V., Leseleuc, S. *et al.* (2018). Synthetic three-dimensional atomic structures assembled atom by atom. *Nature* 561:79–82.

Bednorz, J.G. & Muller, K.A. (1986). Possible high TC superconductivity in the Ba-La-Cu-O system. *Zeit. Phys. B.* 64(2):189–93.

Bennett, C.H. & DiVincenzo, D.P. (2000). Quantum information and computation. *Nature* 404:247–55.

Chen, L., Jordan, S., Liu, Y.-K. *et al.* (2016). Report on post-quantum cryptography. NIST Interagency Report 8105.

Deutsch, D. (1985). Quantum theory, the Church-Turing principle and the universal quantum computer. *Proc. Roy. Soc. Lond. A.* 400(1818):97–117.

DiVincenzo, D.P. (2000). The physical implementation of quantum computation. *Fortschrit. Phys.* 48(9–11):771–83.

Drozdov, A.P., Eremets, M.I., Troyan, I.A. *et al.* (2015). Conventional superconductivity at 203 kelvin at high pressures in the sulfur hydride system. *Nature* 525(7567):73–6.

Dyakonov, M. (2018). The case against quantum computing: The proposed strategy relies on manipulating with high precision an unimaginably huge number of variables. *IEEE Spectr.*

Feynman, R.P. (1960). There's plenty of room at the bottom. *Eng. Sci.* 23(5):22–36.

Feynman, R.P. (1982). Simulating physics with computers. *Int. J. Theor. Phys.* 21(6):467–88.

Freedman, M.H., Kitaev, A., Larsen, M.J. & Wang, Z. (2002). Topological quantum computation. arXiv:quant-ph/0101025.

Gasman, L. (2019). Quantum key distribution (QKD) markets: 2019 to 2028. Inside Quantum Technology Report.

Gidney, C. & Ekera, M. (2019). How to factor 2048 bit RSA integers in 8 hours using 20 million noisy qubits. arXiv:1905.09749 [quant-ph].

Grumbling, E. & Horowitz, M. (2019). *Quantum Computing: Progress and Prospects.* Washington, DC: US National Academies of Sciences.

Haque, A. & Sumaiya, S. (2017). An overview on the formation and processing of nitrogen-vacancy photonic centers in diamond by ion implantation. *J. Manuf. Mater. Process.* 1(1):6.

Jack, B., Xie, Y., Li, J. *et al.* (2019). Observation of a Majorana zero mode in a topologically protected edge channel. *Science* 364(6447):1255–59.

Kadowaki, T. & Nishimori, H. (1998). Quantum annealing in the transverse Ising model. *Phys. Rev. E.* 58(5355).

Kaminsky, W.M. & Lloyd, S. (2004). Scalable architecture for adiabatic quantum computing of Np-hard problems. In: Leggett A.J., Ruggiero B. & Silvestrini P. (Eds). *Quantum Computing and Quantum Bits in Mesoscopic Systems.* Boston, MA: Springer.

Kloeffel, C. & Loss, D. (2013). Prospects for spin-based quantum computing in quantum dots. *Annu. Rev. Conden. Matt. Phys.* 4:51–81.

Levine, H., Keesling, A., Omran, A. *et al.* (2018). High-fidelity control and entanglement of Rydberg-atom qubits. *Phys. Rev. Lett.* 121(123603).

Los Alamos National Laboratory (LANL). (2004). A Quantum Information Science and Technology Roadmap. LA-UR-04-1778.

Loss, D. & DiVincenzo, D.P. (1998). Quantum computation with quantum dots. *Phys. Rev. A.* 57(1):120–26.

Mavroeidis, V., Vishi, K., Zych, M.D. & Josang, A. (2018). The impact of quantum computing on present cryptography. *Int. J. Adv. Comp. Sci. App.* 9(3):1–10.

McMahon, P. (2018). *Quantum Computing Hardware Landscape*. San Jose, CA: QC Ware.

Mosca, M. (2018). Cybersecurity in an era with quantum computers: will we be ready? *IEEE Secur. Priv.* 16(5):38–41.

Murali, P., Linke, M. & Martonosi, M. (2019). *Full-Stack, Real-System Quantum Computer Studies: Architectural Comparisons and Design Insights.* International Symposium on Computer Architecture (ISCA), 2019, pp. 1–14.

Petta, J.R., Johnson, A.C., Taylor, J.M. *et al.* (2005). Coherent manipulation of coupled electron spins in semiconductor quantum dots. *Science* 309(5744): 2180–84.

Preskill, J. (2018). Quantum computing in the NISQ era and beyond. *Quantum* 2(79):1–20.

Procopio, L.M., Moqanaki, A., Araujo, M. *et al.* (2015). Experimental superposition of orders of quantum gates. *Nat. Commun.* 6(7913):1–6.

Rigetti, C., Poletto, S., Gambetta, J.M. *et al.* (2012). Superconducting qubit in waveguide cavity with coherence time approaching 0.1 ms. *Phys. Rev. B.* 86:100506(R).

Robinson, N.J., Altland, A., Egger, R. *et al.* (2019). Nontopological Majorana zero modes in inhomogeneous spin ladders. *Phys. Rev. Lett.* 122(2):027201.

Rudolph, T. (2016). Why I am optimistic about the silicon-photonic route to quantum computing. arXiv:1607.08535 [quant-ph].

Saffman, M. (2016). Quantum computing with atomic qubits and Rydberg interactions: progress and challenges. *J. Phys. B: Atom. Mol. Opt. Phys.* 49(202001):1–27.

Sarma, S.D., Freedman, M. & Nayak, C. (2015). Majorana zero modes and topological quantum computation. *NPJ Quantum Inf.* 1(15001).

Somayazulu, M., Ahart, M., Mishra, A.K. *et al.* (2019). Evidence for superconductivity above 260 K in lanthanum superhydride at megabar pressures. *Phys. Rev. Lett.* 122(027001).

Steane, A. (1997). Quantum computing. arXiv:quant-ph/9708022.

Vandersypen, L.M.K., Steffen, M., Breyta, G. *et al.* (2001). Experimental realization of Shor's quantum factoring algorithm using nuclear magnetic resonance. *Nature* 414:883–7.

Villalonga, B., Boixo, S. & Nelson, B. (2019). A flexible high-performance simulator for the verification and benchmarking of quantum circuits implemented on real hardware. arXiv:1811.09599 [quant-ph].

Wang, Z. (2010). *Topological Quantum Computation.* Providence, RI: American Mathematical Society.

Williams, C.J. (2007). *Quantum Information Science, NIST, and Future Technological Implications.* Gaithersburg, MD: National Institute of Standards and Technology.

Zurek, E. (2019). Viewpoint: pushing towards room-temperature superconductivity. *APS Phys.* 12(1).

Chapter 4

Advanced Quantum Computing: Interference and Entanglement

Abstract

The special properties of quantum objects (atoms, ions, and photons) are superposition, interference, and entanglement. Superposition refers to particles existing across all possible states simultaneously. Interference is the situation where intervention from noise in the environment damages the quantum object, and also the possibility that the wave functions of particles can either reinforce or diminish each other. Entanglement means that groups of particles are connected and can interact in ways such that the quantum state of each particle cannot be described independently of the state of the others even when the particles are separated by a large distance. One of the most important implications of entanglement is that qubits can be error-corrected, which will likely be necessary for the advent of universal quantum computing. An application of quantum computing that is already available is certifiably random bits, a proven source of randomness, which is used in secure cryptography.

4.1 Introduction

One surprise is that there may be many more useful short-term applications of quantum computing with currently available NISQ devices than has been thought possible without full-blown universal quantum computers. NISQ devices are noisy intermediate-scale quantum devices

(Preskill, 2018). For example, even near-term quantum computing devices may allow computations as elaborate as the simulation of quantum field theories (Jordan *et al.*, 2012).

4.1.1 *Quantum statistics*

Quantum superposition, entanglement, and interference (SEI) properties come together in the discipline of quantum statistics. Quantum phenomena have a signature. They produce certain kinds of recognizable quantum statistical distributions that could only have come from quantum mechanical systems. This includes patterns from interference (through amplitudes), superposition (through qubit spins, such as in quantum annealing), and entanglement (through Bell pairs and otherwise). These quantum signatures are unique and identifiable. This is not surprising, given that quantum statistics means studying how wave functions behave in a quantum mechanical system, in a statistical format (i.e. distribution-based). The key point is that only a quantum system could have produced such output, and thus it can be used as a source of provable randomness.

As an indication of the unique signifiers of quantum phenomena, one relevant interpretation of quantum statistics is Porter–Thomas distribution. These are distributions in which the probabilities themselves are exponentially distributed random variables. The quantum statistics are known and have been developed elsewhere in physics to model quantum many-body systems. The practical application is that quantum statistical distributions can be set up to generate either predictable patterns or randomness. In particular, many applications require a guaranteed source of randomness. True randomness instills trust in believing that events have been fairly determined. Some of the immediate applications for randomness are cryptography (setting the parameters for a system that cannot be back-doored or otherwise breached), and blockchains more generally, both in cryptography and in facilitating the creation of next-generation consensus algorithms (PBFT) based on entropy. Other uses for randomness include running lotteries (picking numbers fairly) and auditing election results (selecting precincts to review at random).

At present, most randomness is not guaranteed to be random, and a potential trend would be the widespread use of quantum computers to generate guaranteed randomness for use in various security applications.

4.2 Interference

In physics, wave interference is a phenomenon in which two waves in a system have either a reinforcing or canceling effect upon one another. There can be positive coherence if the two waves are in the same phase, reinforcing each other in a stronger way, as in the ocean when multiple big waves come into shore at once. Alternatively, there can be negative coherence if waves are in phases that counterpose or cancel each other out, such as when there is noise from the environment.

Interference is used in building quantum circuits and calculating with vectors in quantum computing. A quantum circuit harnesses the qubit wave action with matrix multiplications (linear algebra). Each time a vector (corresponding to qubit position) is multiplied by a matrix (the computational movement through the quantum gate system), the matrix combines numbers in the vector, and the combination either reinforces the numbers or cancels them out. In this real physical sense, coherent wave behavior is calculated as a vector passing through quantum gates. This is an important factor that is competing against the fact that coherent wave behavior is the environmental noise of the system. The coherent action of the waves is fragile, and can be easily destroyed if the system has too much noise or other interference.

This is a challenge in quantum computing because irrespective of the qubit-generation method (superconducting circuits, trapped ions, topological matter, etc.), there is always going to be noise in the system, and if the noise overwhelms the coherent wave activity, the quantum computer is not going to work. Hence, quantum error correction becomes important for mitigating the noise.

4.2.1 *Interference and amplitude*

The wave behavior of qubits and interference is seen in modeling coherent wave action through quantum gates (protecting against noise in quantum circuit design), and also in another property of the quantum mechanical domain, amplitude. Whereas probabilities are assigned to the different possible states of the world in classical systems, amplitudes are the analog in quantum systems. Amplitudes are more complicated than probabilities, in that they can interfere destructively and cancel each other out, be complex numbers, and not sum to one. A quantum computer is a device that maintains a state that is a superposition of every configuration of qubits measured in amplitude. For practical computation, the amplitudes are converted into probabilities (probability is the squared absolute value of its amplitude). A key challenge is figuring out how to obtain a quantum speed advantage by exploiting amplitudes. This is not as straightforward as using the superposition property of qubits to model a greater number of possibilities, since simply measuring random configurations will not coalesce into problem-solving answers. Hence, quantum statistical models are implicated to exploit amplitudes such that certain patterns of interference are produced.

To produce the kinds of interference patterns that might be directed into problem answers, one strategy is engaging the properties of wave coherence. In a quantum circuit, each of the amplitudes of the possible output states is the sum of exponentially many possible contributions. These contributions are complex numbers pointing in every direction in the complex plane, and the final amplitude is whatever residue is left over after the complex numbers have mostly collapsed and canceled each other out in the ending state. The idea is to incorporate this model of amplitudes into a quantum-solvable process.

An analogy in the everyday world can be made with light. A beam from a laser pointer could be shone through a field of ground-up glass to see where the light ends up on a screen at the end of the field. This produces a speckling pattern, in that as the beam goes through the field, there are darker points where there is destructive interference, and lighter points where there is constructive interference. Running many samples firmly establishes the pattern of where the individual photon lands. The consistency of the speckle pattern can then be analyzed to see if the photon preferentially lands on the lighter points of the constructive interference or the darker points of the destructive interference.

There are two possible ways the amplitude interference patterns can be used, to produce a reliably repeatable pattern or to produce a random pattern. The first idea is to generate a predictable pattern, which implies that this particular interference system can be used to encode a real-world problem, such that a useful answer can be interpreted. Conceptually, this is more or less how quantum annealing operates, although qubit spins, not interference, is the mechanism. The same principle is at work in encoding a real-world problem into a quantum-solvable process where the quantum system runs and provides an answer. The other possibility is that a predictable pattern is not the outcome, that the result is random, which is helpful in another way. Gaussian output is an indication of randomness, of a well-formed statistically sound mechanism for generating randomness. Overall, the implication of quantum statistics is that quantum randomness can be produced, even in NISQ devices.

4.3 Noisy Intermediate-Scale Quantum Devices

The long-term goal of quantum computing is to realize universal quantum computation on fault-tolerant quantum information processors. In the shorter term, the objective is to solve problems with NISQ devices, which are quantum processors that are not error-corrected. Quantum computing is developing in different steps based on available technical functionality. The first phase of quantum computing (2001–2012) consisted of several demonstrations of 1–2 qubits and up to 10-qubit systems using a variety of hardware approaches. The second phase of quantum computing is currently underway (2012–2019) and includes general-purpose 30–70 qubit systems with gate model superconducting circuits, and special-purpose 2048-qubit systems with quantum annealing machine superconducting circuits. The landmark discovery in 2012 of high-temperature superconductors helped to propel the development of general-purpose gate model logic in superconducting circuits, in which operations can be controlled. Superconducting circuits, whether based in standard gate models or quantum annealing machines, are the only hardware approaches that are commercially available as of June 2019.

Existing quantum computers are NISQ devices, meaning imperfect, modestly sized quantum computing machines (Preskill, 2018). NISQ devices are an important advance over the few-qubit systems that largely served as a proof of concept. The challenge with NISQ devices is finding relevant problems that can be solved with only 50–100 qubits without error correction. In the longer-term, quantum computers are foreseen to have the most significant advantage over classical computers in solving computational problems that are known to be difficult. These include

Shor's factoring algorithm (which could likely break the current RSA cryptography standard) and Grover's search algorithm (for faster search through large datasets). Nevertheless, in the shorter term, even with NISQ devices, it may be possible to make significant progress in the areas of optimization, simulation, machine learning, and cryptography. One application with significant economic promise for quantum computing is simulating physical processes in chemistry, physics, and biology to discover new materials and pharmaceutical substances.

Although error correction is not a substantial issue for NISQ devices, it could start to become a problem in more sophisticated quantum computing systems given the sensitivity of qubits to environmental noise. Predictions as to the number of qubits for which error correction will be required vary considerably. One estimate is presented in Table 4.1 (McMahon, 2018), as a strawman sketch of the different kinds of applications and the number of qubits needed. Many factors could play a role in error correction, including the hardware approach, chip architecture, and quantum algorithm design. Due to the short-term unavailability of quantum error-correction schemes, clever workarounds have been developed. These include using NISQ devices for quantum simulation using error-resilient algorithms (Colless *et al.*, 2018), hardware that is resistant to errors (Kandala *et al.*, 2017), and other error mitigation techniques (Kandala *et al.*, 2019).

One step in the quantum computing roadmap is demonstrating quantum advantage (a clear advantage of using quantum computing versus

Table 4.1. Quantum applications and number of qubits required.

Application	Estimated # qubits required
Quantum advantage	50
Quantum optimization and sampling	~100
Quantum simulation	>100
Chemistry simulation/nitrogen fixation	Few hundred
Applications requiring error correction	>1 million
Shor's algorithm (factoring)	>50 million
Grover's algorithm (search)	>100 million

classical computing). Substantiating quantum advantage is a largely academic hurdle that is expected to be achievable with the NISQ devices that are currently available with 30–70 qubits. After quantum advantage, the next class of quantum computing applications is optimization and simulation in domains ranging from chemistry, physics, and biology to economics and finance. Further applications involve quantum machine learning and quantum cryptography. Quantum machine learning refers to both the application of machine learning techniques to the quantum domain and using quantum mechanical systems to extend research in machine learning. Quantum cryptography contemplates an extensive suite of applications such as quantum key distribution, cryptographic algorithms, and quantum-secure zero-knowledge proofs.

4.3.1 *Computability and computational complexity*

The advent of quantum computing calls into sharper relief theories for analyzing the kinds of problems that are computable. Computability relates to the theory of computation, and investigates how efficiently problems can be solved based on different kinds of algorithms and models of computation. Quantum computing might extend the range of the kinds of problems that can be computed efficiently, but would not allow the computability of all problems. Quantum computing could provide an incremental yet crucial extension to the kinds of real-world problems that might be solvable.

In computational complexity, two parameters are studied, the required computation resources in terms of time and space that are necessary to calculate the problem. The theoretical basis for computational complexity is the Church–Turing computability thesis (Table 4.2). The original Church–Turing thesis is concerned only with the theoretical status of whether a given problem is computable at all, irrespective of the time required. The extended Church–Turing thesis also incorporates the practical consideration of whether a problem is efficiently computable in time. A given problem might fall within any of various tiers of time complexity and space complexity in the hierarchy of computational complexity classes.

As an extremely broad heuristic, quantum computers may allow a one-tier increase in computing according to the computational complexity

Table 4.2. Church–Turing computability thesis.

Church–Turing thesis	Problem answered
Church–Turing (original)	Is the problem computable? (ignoring time)
Church–Turing (extended)	Is the problem efficiently computable in time?

schema. For example, a problem that requires exponential time in classical systems (i.e. time that is too long for practical results) may take polynomial time in quantum systems (i.e. a reasonable amount of time for practical use). In the canonical Traveling Salesman Problem, it may be possible to check twice as many cities in half the time using a quantum computer.

4.4 Quantum Error Correction

4.4.1 *Practical concerns and status*

One of the biggest challenges to the potential instantiation of universal quantum computers is quantum error correction. Qubits are more fragile than classical bits and need to be error-corrected if they become damaged. It is too early in the development of quantum computing to estimate exactly which hardware approaches will require error correction and at what point (with certain numbers of qubits). However, since qubits are affected by both state decay and environmental noise, some kind of error correction is likely to be required. Methods for error correction are largely a research effort at present, and more clarity may arrive with implementation.

The two kinds of commercially available systems (as of June 2019) use superconducting circuits, but in different architectures. There is the standard gate model (with 30–70 qubits) and the quantum annealing model (with 2048 qubits). The trade-off is that the gate model aims to be more of a universal quantum computer, whereas the quantum annealing model can only accommodate certain classes of optimization problems. On the one hand, annealing machines harness the natural evolution of quantum systems over time to settle into the lowest-energy state (a minimum). The intermediate operation of the quantum process cannot be

Table 4.3. Quantum computing systems and error correction.

System	Error correction required?	Number of qubits demonstrated	Qubit composition	System status
Superconducting: Standard gate model	Yes (at >1 million qubits)	30–70	Matter	Commercially available
Superconducting: Quantum annealing	Not initially	2048	Matter	Commercially available
Ion trapping	Not initially	27	Matter	Research
Majarana fermions	Not initially	Unknown	Matter	Research
Photonic quantum circuits	Minimal	Unknown	Light	Research

controlled. On the other hand, gate model machines seek to control and manipulate the evolution of the quantum system at each moment in time, which suggests that they can be used to solve a much larger and more general set of problems. For error correction, the benefit is that with less manipulation, less errors arise, and the annealing machine does not require the same kinds of error correction that the gate model machine does, at least at the current number of qubits.

Other approaches to quantum information systems claim that error correction is not an immediate concern (Table 4.3), although the path to scaling up the number of qubits is perhaps less clear than with superconducting circuits. These other methods, although still in the research phase, are designed such that error-correction requirements may be greatly reduced. These systems include ion trapping, fermion braiding, and photonic quantum computing. In particular, the fermion braiding approach is robust to noise since changes in geometry do not have an effect, only changes in topology, and the errors can be corrected in hardware instead of software.

4.4.2 *Quantum state decoherence*

Quantum information has different properties than classical information and is much more sensitive to being damaged by the computing environment. Whereas in a classical system, the idea is to pack in as many bits as possible, in a quantum system, the aim is to have only as many high-fidelity qubits as can be effectively controlled, and scale up with that level of integrity. It is difficult to isolate systems from the environment well enough for them to have useful quantum behavior. The quantum states can decohere (decay) quickly and become damaged by the noisy (imperfect) environment. It is likely that some kinds of error correction will always be necessary in quantum systems. Even if a perfect computing environment without any noise were to be obtained, there is still the natural property of qubits to decay that must be addressed. The excited state of qubits ultimately decays to the ground state due to being coupled to the vacuum fluctuation (the vacuum fluctuations of electromagnetic fields are unavoidable). Therefore, it is necessary to encode the quantum states to be protected in such a way that they can be error-corrected to remain robust. These requirements introduce the notions of qubit lifecycle and qubit management in quantum information systems.

4.4.3 *Entanglement property of qubits*

Unlike classical information, which can be examined arbitrarily many times to determine if it has changed, quantum information cannot be measured directly because it will change or destroy the information. However, quantum information has the interesting property of entanglement. The entanglement property refers to quantum particles being entangled with one another. Quantum particles are not isolated and discrete, but rather correlated, both with each other, and the history of the system. Particles and previous states are correlated as part of the fuller information landscape of the quantum system. Hence, quantum error correction is performed by taking advantage of the entanglement property of qubits. The insight is that due to entanglement, it is possible to measure relationships between qubits without measuring the values stored by the qubits themselves.

4.4.3.1 *Error-correction codes*

Quantum error correction uses the entanglement property to smear out the information of 1 qubit onto 9 entangled qubits (in the basic case). The idea is that the information of 1 qubit can be stored on a highly entangled state of several qubits. A local point, 1 qubit, can be smeared out, and represented with a larger number of qubits. The auxiliary qubits are called an ancilla (ancillary qubits). Quantum error correction is typically performed by introducing an ancilla in the form of additional qubits with which the qubit of interest is entangled. Entangling the qubit of interest with ancillary qubits allows the qubit to be protected by smearing out its information over a larger system. The qubit of interest can be error-checked since its information can be examined indirectly through the entangled qubits (through parity measurements). In this way, quantum error correction is used to test the integrity of quantum information without damaging it. Quantum error correction makes quantum computing feasible, in that a quantum computer can tolerate some degree of noise in the environment by correcting errors. Many different kinds of quantum error-correction codes (encoding schemes) have been proposed. Shor's code uses a 9-qubit smear, and others require fewer qubits (a 7-qubit code and a 5-qubit code, for example) (Shor, 1995).

4.4.3.2 *Classical error correction*

In general, all forms of computing systems use error correction to check for data integrity. An error-correcting process seeks to determine whether information has been damaged or destroyed and restores its initial state. Error correction is well-understood in classical logic. For example, there could be a memory chip storing bits that is hit by a cosmic ray. If one bit in a 32-bit word is flipped, there are many known ways of recovery. One frequently used method is having many copies of the data. With redundancy, having several copies of the information means that a mechanistic majority-voting mechanism can be used to confirm the intact version of the data. With quantum logic, however, error correction based on making redundant copies of the same information cannot work. It is not possible to copy quantum information due to the no-cloning theorem, which states that it is impossible to create an identical copy of an arbitrary unknown quantum state. Therefore, quantum error-correction methods such as those based on entanglement are needed.

4.4.3.3 *Shor's code*

It is not by accident that Shor's code, the first quantum error-correction code discovered, is 9 qubits. Nine qubits is the smallest ancilla that can be used to confirm that the original qubit was not flipped (changed or damaged), by checking various pairwise sequences of possible flipping along the *X*, *Y*, and *Z* axes of the qubit. A simple error-correcting code could instantiate a single logical qubit of data as three physical qubits for each scenario of the three axes. With pair-wise evaluation, it is possible to determine whether the first and second qubits have the same value, and whether the second and third qubits have the same value, without determining what that value is. If one of the qubits turns out to disagree with the other two, it can be reset to their value. Further, the pair-wise evaluations might be performed in both time and space suggesting quantum information processing architectures with time speed-ups.

Shor's code is a sequential method using single Pauli operators (3D operators with *X*, *Y*, and *Z* values) to act on the system according to the different possible error permutations that could have occurred. Since the ancillary qubits and the original qubit are entangled, any error will have a recognizable signature and can be corrected (by repairing it into the initial phase or into an irrelevant phase that does not impact the original qubit's information). Since the states of the system are eigenstates of eigenvalue one for all of the operators, the measurement does nothing to the overall state. The Shor code is redundant, in that the number of bits of information it protects is significantly fewer than the number of physical bits that are present. Noise can come in from the environment without disrupting the message. Also, the Shor code is nonlocal in the sense that the qubit information is carried in the entanglement between the multiple qubits that protect it against local decoherence and depolarization.

The Pauli operator that uses *X–Y–Z* quantum spin representations is one proposed method. Kraus operators, which are operator-sum representations, are another (Verlinde & Verlinde, 2013). However, Kraus operators can be difficult to engage because they require details of the interaction with the environment, which may be unknown. Single Pauli operators suggest a more straightforward implementation model.

4.4.4 *Quantum information processors*

The main concept of error correction is that to protect qubits from environmental noise and to mitigate against state decay, the qubit of interest can be encoded in a larger number of ancillary qubits through entanglement. The entangled qubits are combined into a bigger overall fabric of qubits that constitutes the quantum information processor. For example, a quantum information processor might have 50 qubits, and an error-correction requirement of 9 qubits for each qubit. This would only leave 5 qubits available for information processing. The scaling challenge is clear, in that with current error-correction methods, most of the processing capacity must be devoted to error correction. The implied scaling rule is 10, in the sense that any size quantum information processor only has available one-tenth of the total qubits for the actual information processing (each 1 qubit requires 9 qubits of error correction). More efficient error-correcting codes have been proposed, but the scaling rule of 10 could be a general heuristic, at least initially.

4.5 Bell Inequalities and Quantum Computing

4.5.1 *Introduction to inequalities*

Mathematical inequalities are important concepts in quantum computing and statistical physics. A mathematical inequality is two quantities that are not equal to one another. More formally, an inequality is the situation in which two quantities have different values (an inequality contrasts with an equality, in which two quantities have the same value).

The theoretical implication of mathematical inequalities is that they can be used as a technique to prove other mathematical statements, or make statements about other mathematical statements that might be useful in problem solving. For example, inequalities can be used to frame problems by bounding quantities for which exact formulas cannot be easily computed. As such, mathematical inequalities can be used as an analytic tool to marshal probabilistic data, in various application areas such as machine learning and quantum mechanical systems. The idea is using mathematics as a tool to rewrite or transform complicated problems into computable problems, and then solve them and prove things about them. There are hundreds of different mathematical inequalities. Most relevant to quantum computing and statistical physics are Chebyshev, Jensen, Cauchy–Schwarz, and Bell inequalities.

4.5.1.1 *Chebyshev's inequality*

One way that inequalities are used is as a tool for analyzing probabilities in large datasets. In the practical case of automobile insurance, claims are independent events that occur at random times and in random sizes. To estimate required capital reserves, it is useful to have a function indicating the distribution and predicting the maximum loss. Chebyshev's inequality is used to project the event distribution of insurance claims, and to structure the distribution such that 90% of future claims would be expected to fit certain metrics such as falling within three standard deviations of the mean.

4.5.1.2 *Jensen's inequality*

Another way that inequalities are used is to determine convex and concave functions. A convex function is an upward-facing curve (an empty bowl) and a concave function is a downward-facing curve (an upside-down bowl). Having convex and concave functions is useful for optimization because the tools of calculus can be applied to identify minima and maxima. Defining a phenomenon in the profile of an s-curve, with convex and concave portions, suggests the possibility of risk management to accentuate positive outcomes and limit negative outcomes. Jensen's inequality is used in financial markets (Taleb, 2007) and medicine (for optimal patient drug dosing) (Reynolds, 2010).

4.5.1.3 *Cauchy–Schwarz inequality*

The Cauchy–Schwarz inequality is important because it has been used to justify the idea that Hilbert spaces are generalizations of the Euclidean space. Although the most basic Hilbert space is a 3D space like the everyday macroscale world (Euclidean space), more formally, Hilbert spaces are arbitrarily many dimensional spaces that do not exist in the physical world. Mathematician David Hilbert invented the notion of Hilbert space as a structure for reasoning about abstract mathematical objects beyond those that are well-determined (Dieudonné, 1983, pp. 115–116).

The Cauchy–Schwartz inequality is used to prove that Hilbert space is a generalization of Euclidean space by demonstrating the convergence of a sequence of vectors in inner product space, and also the continuity of the inner product space. The Cauchy–Schwarz inequality relates vectors of inner products to one another with equality or inequality based on certain conditions (specifically, relating the magnitude of a vector to its inner product with another vector with equality if the vectors are linearly dependent (Kwan & Greenstreet, 2018, p. 111). This inequality formulation is relevant to concepts such as defining the surface in the bulk/boundary correspondence as a Cauchy surface with a time dimension, and therefore dynamics.

4.5.2 *Bell inequalities*

In 1964, physicist John Bell developed Bell's theorem (the claim that particles in a quantum system remain dependent, no matter how far apart they are) and Bell's inequality as a test to disprove the hidden variables theory (Bell, 1964). The crux of Bell's experiment is noticing that the two different arms of the EPR paradox would deliver different predictions, and that these could be tested mathematically with inequalities. The most crucial inequality for quantum computing is Bell's inequality (also called Bell inequalities), in particular, the point that Bell's inequality cannot be violated, which allows entangled quantum pairs to be produced. Bell's inequality is implicated in resolving the EPR paradox. Proposed in 1935, the Einstein–Podolsky–Rosen (EPR) paradox is a thought experiment proposed by Einstein, Podolsky, and Rosen to demonstrate what they thought was a lack of completeness in quantum mechanics (Einstein *et al.*, 1935). The paradox refers to the problem that in a system with two entangled particles, if one particle is measured, the other particle is affected, no matter how far apart they are. The paradox seems to be that one particle's changing must be due either to local variables (instructions carried by the particle or found in the environment) that produce the change, or to information being sent from the other particle, in time that implies faster than speed of light travel, which produces the change.

Of these two possible answers to the paradox, Einstein did not accept that there could be "spooky action at a distance" that would suggest faster than lightspeed travel since nothing can travel faster than the speed of light. Thus, to explain how particles influence one another, EPR suggested the existence of local hidden variables in quantum mechanical systems. Niels Bohr, one of the founders of quantum mechanics, disagreed, and maintained that there are no hidden variables (the Copenhagen interpretation of quantum mechanics).

The predictions made by local hidden variable theories would be different from the predictions made by so-called "spooky action at a distance" theories. Bell set up an inequality based on the opposing values of a system determined by local variables and a system not determined by local variables.

Bell structured equations to indicate that if there are physical properties (local hidden variables) accounting for the measurements in a system, then the inequality is true. A 3D physical system has measures of photon spins along three axes (X, Y, and Z). The basic idea of the inequality is that in such a physical system, a measure of less information would be smaller than a measure of more information. The inequality consists of a measure of the spin of photon pairs that are $X+$ and $Y+$ (i.e. less information), and says that this value would be less than the measure of $X+Z+$ plus Y–Z– (i.e. more information) (Baez, 1996). Bell's inequality is that in a physical system with local information, a measure of less information is not equal to a measure of more information.

Bell showed that if a local hidden variable theory is true, then measurements would have to satisfy certain constraints, called Bell inequalities (Bell inequalities would be true). Likewise, the opposite holds, if Bell's inequalities were violated, hidden value theory would be false (Table 4.4).

Bell's inequality shows that only if the inequality were satisfied could there be local hidden variables. Violating Bell's inequality indicates that there is some other kind of explanation for how remote particles can affect one another. The answer is not "spooky action at a distance" based on faster than light travel, but rather quantum entanglement. Quantum systems have different properties than macroscale systems and Bell's inequality helps to demonstrate this. Bell's inequality shows that there is

Table 4.4. Interpretations of Bell's inequality.

State of Bell's inequality	Theory supported	Status
Bell's inequality is true	There are local hidden variables (EPR)	FALSE
Bell's inequality is violated	There is quantum entanglement	TRUE

a limit to the particle correlations that could be the result of local conditions. Any additional correlations beyond those limits would require either sending signals faster than the speed of light (currently thought to be impossible), or another mechanism such as quantum entanglement.

The advance of Bell's inequality is invalidating any claim of there being local hidden variables as EPR proposed and more generally supporting the understanding that quantum particles remain entangled across large distances. Knowing that Bell's inequality cannot be violated and that quantum particles are entangled is a useful tool for constructing quantum mechanical systems such as quantum computers. The important implication of Bell's inequality is that it proves the entanglement of particles in quantum systems.

Numerous experiments have conclusively demonstrated that the remote entanglement of paired quantum particles is true. Particles are generated together as an entangled pair and then separated across the globe before one is measured (Weihs *et al.*, 1998). What Einstein called "spooky action at a distance" is in fact real (although due to entanglement not faster than lightspeed travel). A further advance, tests for Bell's inequality by measuring the photon polarization of two entangled particles have also been demonstrated experimentally (Shalm *et al.*, 2015; Giustina *et al.*, 2015). This work is deployed in practical settings to create entangled Bell pairs that are used to produce and certify randomness, for example, for use in cryptography.

Bell inequalities cannot be violated in quantum mechanical systems. However, Bell inequalities also arise in macroscale systems, where they can be violated. This is a known situation and gives even more credence to the notion of quantum structure that is genuine and distinct. One example of the Bell inequality macroscale violation is illustrated using connected vessels of water (Aerts *et al.*, 2000). Another example of the

Bell inequality macroscale violation is seen in convex geometry, in the hull problem. Enumerating the points (facets) on a boat's hull produces Bell inequalities which are violated in quantum probability distributions (Peres, 1999). Results from convex geometry are being extended from the class of convex bodies in geometry to classes of functions in mathematics (Li, 2018). Methods in mathematical geometry are in turn being deployed to model quantum mechanical systems, for example, testing facet inequalities in a Hilbert space of increasing dimensionality (Das *et al.*, 2018).

4.6 Practical Applications of Entanglement: NIST Randomness Beacon

The important practical upshot of resolving the EPR paradox by indicating that Bell's inequality cannot be violated is that particle entanglement (so-called quantum nonlocality) is a real phenomenon that can be harnessed in the development of quantum systems. Specifically, the entanglement property of quantum mechanical systems can be used as a source of randomness (entropy).

Bell inequalities can be used to generate certifiably random bits. Randomness is fundamental to the security of digital networks and cryptographic systems. Random numbers are used hundreds of billions of times a day to encrypt data in electronic networks (for example, for credit card authorizations). A prominent provider of randomness is the NIST Randomness Beacon, which generates random bits every 60 s. However, at present, these numbers are not certifiably random. The problem with classical random-number generators is that they are not post-quantum secure and it is hard to ensure that the outputs are unpredictable (truly random). Hence, the practical deployment of quantum randomness is an objective. Quantum randomness has been demonstrated experimentally by several teams, through generating entangled Bell pairs (Shalm *et al.*, 2015; Giustina *et al.*, 2015). The demonstrations indicate that it is possible to exploit the phenomenon of quantum nonlocality with a Bell test to build a random-number generator that can produce an output that is truly random. The demonstrations are underpinned by theoretical work suggesting that a strong form of randomness generation is impossible classically, and would be possible in quantum systems only if certified by a Bell inequality violation (Pironio *et al.*, 2010).

4.6.1 *Certifiably random bits*

A recent advance is the certifiably random bit, meaning certifying that the randomness has been produced as a result of quantum mechanics and not some other method (Bierhorst *et al.*, 2018). Whereas in classical computing, it is difficult to prove that random bits are truly random and unpredictable, in quantum computing, the method for generating bits can be guaranteed to be random through quantum mechanical principles. This is because the kinds of statistical distributions and correlations obtained between measurement and outcome could only be the result of quantum computation.

A so-called loophole-free Bell test demonstrates the production of certifiably random bits. Bits are loophole-free in the sense that they could not have been predicted according to any physical theory and must have been quantum generated. Bits are generated with photons, entangled, and measured for randomness. Entanglement is produced by using a laser to hit a crystal that converts light into pairs of polarization-entangled photons. During the entanglement phase of the process, a long string of bits is generated through a Bell test. Correlations are measured between the properties of pairs of photons. Statistical tests of the correlations prove that the bits have been generated by a quantum system, and also allow the amount of randomness to be quantified. In this particular experiment, the Bell pair test data indicated randomness being spread very thinly throughout the long string of bits, with nearly every bit being 0 and only a few being 1. Therefore, an extraction was performed to obtain a short, uniform string with concentrated randomness such that each bit has a 50/50 chance of being 0 or 1.

The result is that the final bits are certified to be random. The randomness is certifiable because given the constraints of independent statistical measurement principles and no faster-than-lightspeed travel, only a quantum system could have produced this kind of statistical output. The ability to certify and quantify randomness is an important step in developing and implementing quantum systems. To improve the ease of quantum randomness generation, another work proposes that it is possible to produce certifiable randomness within only one quantum device (Brakerski *et al.*, 2018). More closely related to quantum computing, another research demonstrates Bell's inequality violations with remotely connected superconducting qubits (photonic qubits) (Zhong *et al.*, 2019). The work is notable as the first demonstration of Bell's inequality violations in a superconducting system.

References

Aerts, D., Aerts, S., Broekaert, J. & Gabora, L. (2000). The violation of bell inequalities in the macroworld. *Found. Phys.* 30(9):1387–414.

Baez, J. (1996). Does Bell's inequality rule out local theories of quantum mechanics? http://math.ucr.edu/home/baez/physics/Quantum/bells_inequality.html. Accessed June 30, 2019.

Bell, J.S. (1964). On the Einstein-Podolsky-Rosen paradox. *Physics.* 1(3):195–200.

Bierhorst, P., Knill, E., Glancy, S. *et al.* (2018). Experimentally generated randomness certified by the impossibility of superluminal signals. *Nature* 556:223–6.

Brakerski, Z., Christiano, P., Mahadev, U. *et al.* (2018). A cryptographic test of quantumness and certifiable randomness from a single quantum device. arXiv:1804.00640 [quant-ph].

Colless, J.I., Ramasesh, V.V., Dahlen, D. *et al.* (2018). Computation of molecular spectra on a quantum processor with an error-resilient algorithm. *Phys. Rev.* X8:011021.

Das, A., Datta, C. & Agrawal, P. (2018). New facet Bell inequalities for multi-qubit states. arXiv:1809.05727 [quant-ph].

Dieudonné, J. (1983). *History of Functional Analysis*, Vol. 49. Amsterdam, NL: North Holland.

Einstein, A., Podolsky, B. & Rosen, N. (1935). Can quantum mechanical description of physical reality be considered complete? *Phys. Rev.* 41:777–80.

Giustina, M., Versteegh, M.A.M., Wengerowsky, S. *et al.* (2015). Significant-loophole-free test of Bell's theorem with entangled photons. *Phys. Rev. Lett.* 115:250401.

Jordan, S.P., Lee, K.S.M. & Preskill, J. (2012). Quantum algorithms for quantum field theories. *Science* 336:1130–3.

Kandala, A., Mezzacapo, A., Temme, K. *et al.* (2017). Hardware-efficient variational quantum eigensolver for small molecules and quantum magnets. *Nature* 549:242–6.

Kandala, A., Temme, K., Corcoles, A.D. *et al.* (2019). Error mitigation extends the computational reach of a noisy quantum processor. *Nature* 567:491–5.

Kwan, C. & Greenstreet, M.R. (2018). Real vector spaces and the Cauchy–Schwarz inequality in ACL2(r). *EPTCS* 280:111–27.

Li, B. (2018). Convex Analysis and its Application to Quantum Information Theory. PhD Thesis. Case Western Reserve University.

McMahon, P. (2018). *Quantum Computing Hardware Landscape*. San Jose, CA: QC Ware.

Peres, A. (1999). All the Bell inequalities. *Found. Phys*. 29:589–614.

Pironio, S., Acín, A., Massar, S. *et al.* (2010). Random numbers certified by Bell's theorem. *Nature* 464:1021–4.

Preskill, J. (2018). Quantum computing in the NISQ era and beyond. *Quantum.* 2(79):1–20.

Reynolds, A.R. (2010). Potential relevance of bell-shaped and u-shaped dose-responses for the therapeutic targeting of angiogenesis in cancer. *Dose Response*. 8(3):253–84.

Shalm, L.K., Meyer-Scott, E., Christensen, B.G. *et al.* (2015). A strong loophole-free test of local realism. *Phys. Rev. Lett.* 115(250402).

Shor, P.W. (1995). Scheme for reducing decoherence in quantum computer memory. *Phys. Rev. A* 52:R2493(R).

Taleb, N.N. (2007). *The Black Swan: The Impact of the Highly Improbable*. New York, NY: Random House.

Verlinde, E. & Verlinde, H. (2013). Black hole entanglement and quantum Error correction. *J. High Energ. Phys*. 2013:107.

Weihs, G., Jennewein, T., Simon, C. *et al.* (1998). Violation of Bell's inequality under strict Einstein locality conditions. *Phys. Rev. Lett.* 81(23):5039.

Zhong, Y.P., Chang, H.-S., Satzinger, J. *et al.* (2019). Violating Bell's inequality with remotely connected superconducting qubits. *Nat. Phys*. 15:741–4.

Part 2

Blockchain and Zero-Knowledge Proofs

Chapter 5

Classical Blockchain

Abstract

Many developments are underway in blockchains. An overall theme is that a new level of complexity is being abstracted into digital economic networks to accommodate financial transactions beyond payments. The migration to blockchain-based systems appears as a replay of the entirety of world economic history in the space of a few years. Both public and private blockchains are instantiating PrivacyTech and ProofTech through zero-knowledge proof technology, verifiable computing, and user-centric digital identity solutions. There are efficiency, privacy, and security upgrades planned for the large public blockchains (Bitcoin and Ethereum). Layer 2 expansions continue, involving sidechains, payment channels, and off-chain processing to expand the real-time ease of using blockchains. There are new innovations such as stablecoins, smart routing, and next-generation consensus algorithms.

5.1 Introduction: Functionality and Scalability Upgrades

A number of security, scalability, and privacy upgrades are planned for Bitcoin, the largest public blockchain. One focal point is signatures, which are stored in the blockchain and take up a substantial portion of each transaction record. Various methods for compressing signatures have been proposed. The segregated witness (SegWit) plan offloads parts of the signature to a sidechain that is linked to the main chain (thereby segregating the witnessing function). In SegWit, the signature is no longer used to calculate the transaction ID (but is still used in the overall transaction confirmation process). A problem is fixed called transaction malleability, which is the possibility of submitting a different transaction ID with the same spendable amount by changing the digital signature used to create it. Signature protocols are also being upgraded, namely by adding Schnorr signatures as an improvement to the current ECDSA method. Among other features, Schnorr signatures allow greater flexibility in multi-signature transactions (allowing various M of N subsets to sign the transaction; such as 3 of 5 for some transactions, and 4 of 5 for others). Other features are being incorporated in the Bitcoin upgrade such as Merkle trees. This allows for greater flexibility in transaction path-addressing so that a transaction can be sent to a script instead of only a single address, and thus used in a wider range of smart contract functionality.

5.2 Computational Verification and Selectable Trust Models

A key feature of blockchains is that they are trustless. Trust is conferred by the computational system, which removes the need to trust any parties involved with the transaction. So far, the prominent trust model used by blockchains is mining (proof-of-work or proof-of-stake mining). However, more than one computational trust model is possible (Table 5.1). In the future, the trust model could be a user-selected parameter of any transaction. Not every blockchain computation requires the full and expensive security of mining. One conjecture is that if transactions can perform their own proofs, perhaps no mining is needed in certain blockchains. Or, at least mining could be significantly curtailed to serving as even more of an administrative function, only being engaged to batch transactions on a periodic basis for record-keeping.

The trust models listed in the figure demonstrate the Verifier's Dilemma, which is the friction between the economic incentives of the different constituencies involved in each model. The biggest trade-off in blockchains is between keeping the blockchain secure and managing the cost of the security. On the one hand, the rewards for the miners must

Table 5.1. Computational trust model comparison and progression.

Trust model	Trust mechanism
Trusted party	Amazon, bank (centralized intermediary)
Trusted party	Smart contract address (decentralized intermediary)
Trusted majority	Miners: the majority of the participants in a blockchain protocol
Trusted sidechain, off-chain	Batch transactions into the main blockchain trust model (mining)
Trusted multi-signature	M of N refereed participants (possibly selected at random)
Trusted peers	Peer-based mining: peers must confirm two other transactions before submitting their own transaction (IOTA)
Trusted law of large numbers	PBFT and entropy
Trusted proof protocols	SNARKs, bulletproofs, STARKs, Zether

be sufficient for them to be interested in keeping the blockchain secure. On the other hand, if the transaction cost of using the computational platform is too high (transactions fees go to pay the miners), then users are deterred from using the platform. There are other factors at play too, such as overall participant beliefs in the future prospects of the particular blockchain and the issuance of the money supply. (In Bitcoin, the majority of miner rewards are currently derived from the minting of new coins, but this proportion is calculated to decline over time and transaction fees would theoretically become more dominant by the time the money supply is fully outstanding as long as the currency continues to grow in use.)

Ethereum is a trust market and a computation market that is indicative of the Verifier's Dilemma. Any end user can delegate a computation to Ethereum in the form of submitting a transaction or a smart contract. The trust model is that 51% of the network executes the computation and agrees on the answer. The end user asks the entire Ethereum community (all Ethereum miners) to run the computation, which is expensive and unnecessary for many kinds of computations that still need to be computationally verified.

Mining is a demonstrated model of computational trust, but there are others. One such alternative structure of trust models is sidechains, in which the heavy computation is offloaded to sidechains that are linked to the main chains. Another alternative is engaging unknown peers that are randomly selected to perform small tasks in the cooperative ecosystem of blockchains. Examples of this model include trusted random multi-signature (a certain number of peer nodes selected at random confirm a transaction block), or performing a certain behavior in order to access the network (such as confirming two other randomly assigned transactions). Another class of alternative computational trust models is derived through the law of large numbers in next-generation consensus algorithms such as PBFT (Practical Byzantine Fault Tolerance) which uses entropy to randomly select a group of peer signers such as 250 of 300 for each transaction block. Proofs are a new class of computational trust models, namely zero-knowledge proofs in the form of SNARKs, bulletproofs, STARKs, and Zether. Ultimately, the end game could be obtaining trust almost exclusively from cryptographic proofs (including proofs in which the prover is not even required as in quantum proofs).

5.3 Layer 2 and the Lightning Network

Layer 2 refers to protocols that run on top of existing blockchains such as Bitcoin and Ethereum. The aims of Layer 2 solutions are real-time transactions (not needing to wait ten minutes for a confirmation as is the case with Bitcoin), scalability (to have blockchain transactions-per-second metrics that are in line with those of traditional payment networks such as Visa), security, and extended functionality. All manner of next-generation solutions are included in the term "Layer 2" as a catchall. These projects might include sidechains and payment channels, and new functionality for wallets and smart contracts. One Layer 2 project, emblematic in adding new categories of functionality, is smart contract interoperability and another is contracts on contracts (for example, KittyHats and KittyBattles contracts from DapperLabs that run on top of the successful CryptoKitties contracts). One of the most advanced Layer 2 projects is payment channels, specifically as instantiated in the Lightning Network.

5.3.1 *Introduction to the Lightning Network*

Envisioned as a Bitcoin scalability solution, the Lightning Network is a decentralized system in which transactions are sent over a network of micropayment channels (Poon & Dryja, 2016). The transfer of value occurs off-chain through untrusted parties along the network transfer route by contracts which, in the event of uncooperative or hostile participants, are enforceable via broadcast over the blockchain and through a series of decrementing timelocks. The Lightning Network uses onion-style routing such that the payment information and other transaction details are encrypted in a nested fashion so that intermediary nodes along the route only know the previous and next hop in the route.

The benefits of the Lightning Network are that real-time Bitcoin transactions are possible for the first time; transactions that have instantaneous settlement without the ~10-min wait for confirmation. Very small micropayment transactions are free (no transaction fee). The payment channel structure further enables more sophisticated ongoing contractual relationships between parties. Other variations of payment channel solutions have been proposed, the Lightning Network is the most prominent live project.

As of June 30, 2019, the public-facing Lightning Network consisted of 34,242 public channels on 8,944 nodes with a network capacity of 940 Bitcoin (US$10,428,752) (1ML, 2019). The biggest nodes are associated with blockchain ecommerce and retail service providers. Examples include Bitrefill (cryptocurrency gift cards), Blockstream (blockchain financial services), tippin.me (a custodial wallet for receiving and cashing out Lightning payments), Hodl Monkey (crypto clothing (T-shirts and gear)), and Living Room of Satoshi (everyday bill pay with Bitcoin). Consumer-friendly applications are available, for example the Breez Lightning-powered Bitcoin payments app for iPhone launched in June 2019.

Wumbology (the study of being maximal as opposed to minimal) refers to a planned expansion of the channel size and payment size limits that currently exist in the Lightning Network. The limits are intended to protect users from engaging in too much activity on the network due to the early stage of the technology. The current channel size limit is

0.16 Bitcoin and the payment limit is 0.04 Bitcoin (US$1,600 and US$400, respectively, with a Bitcoin price of US$10,000).

Over time, it is possible that a significant portion of cryptocurrency transactions could take place in Layer 2 solutions such as payment channels. Each transaction could include the transaction itself, together with wallet-automated aspects of channel management, rebalancing, and capital planning. The wallet could become the nexus of activities from the end-user perspective. Wallets are evolving in complexity, from only having the capability of building a transaction from one address to another, to payment channel wallets that constitute a sophisticated routing device which can make decisions about capacity allocation, channel optimization, transaction fees, and other tactics.

5.3.2 *Basic routing on the Lightning Network*

The Lightning Network operates (transparently to the end user) by setting up a temporary line of trust between the sender and the recipient with promises that flow forward and receipts that flow backward across the network of Lightning nodes. The payment is transferred in the forward direction through a series of promises between the sender, intermediary routing nodes, and the recipient. A chain of receipts is transferred backwards to confirm the payment and pay any network fees associated with the transaction.

Technically, the sender (the sender's Lightning wallet) sets up a series of promises, one to each intermediary node in the route and one to the final node (the recipient), flowing forward through a number of intermediary nodes to the recipient. When the recipient receives the final promise, the recipient applies the hash to redeem the payment (the hash is the hash of a cryptographic secret included in the initial invoice from the recipient to the sender). The recipient's redemption of the payment triggers a receipt (which also has a copy of the hash) to flow back one hop to the last intermediary node, who applies the hash to redeem their micropayment. This in turn triggers a receipt released to the last-previous intermediary node to redeem their micropayment, and so forth back through the network to the sender. There could be any number of user-selected intermediary nodes, there are typically three nodes involved in standard onion routing.

The Lightning Network uses a combination of source routing and onion routing. Source routing is a standard internet routing protocol in which the sender specifies the route to the recipient. This contrasts with conventional routing, in which routers in the network determine the path incrementally based on the destination. Onion routing is a privacy-protection technique such that each hop only knows the previous hop and the next hop. The initial message has nested layers of instructions for routing the whole path (hence the name onion routing). The first router peels off the outside message layer and executes the instructions which culminate in the information being routed to the next router. That router peels off the then-top message layer and runs the instructions, and so on, until the information payload reaches its destination. Key-based encryption prevents the routers from seeing inside the nested messages.

5.3.3 *Smart routing: Sphinx routing and rendez-vous routing*

The Lightning Network is a bit counterintuitive because it is not possible to send a payment directly to another party, the receiving party must first send an invoice to the sending party. The sender uses the information in the invoice to route the payment, such as the amount, the pay-to address (the recipient's Lightning node public key), and a hash. The hash is the hash of a cryptographic secret that allows each participant in the route to redeem the previous payment and forward the next payment. Using this information, the sender constructs a route to the destination.

New smart routing protocols, sphinx routing and rendez-vous routing, previously developed in computer science, are proposed to eliminate the invoice requirement and more generally improve the usability and security of Lightning transactions. Sphinx routing (connoting unknown enigma, or unknown route) is a cryptographic message format that is essentially an improved onion-like method for relaying anonymized messages. Sphinx routing has additional security features such as hidden path length and the inability to link the legs of a message's journey through a network (Danezis & Goldberg, 2009).

In the Lightning Network, sphinx routing is used to reduce some of the invoicing requirements. The line-of-trust structure still operates the same way, the recipient sends the sender the hash and the routing information, but the recipient does not need to specify an exact amount for the payment. This makes the channel operations more flexible.

Another smart routing advance is rendez-vous routing, a security improvement in which the two parties build the route together (Valloppillil & Ross, 1998). Instead of the sender transmitting all the way to the recipient (and thus knowing information about the recipient), a rendez-vous point is specified at an intermediary node in the network. Using onion routing, the recipient provides the sender with two-step delivery instructions, the first route to the intermediary node and a sealed message that the sender cannot open. The sender delivers the sealed message from the recipient to the specified intermediary node (a public node willing to act as an intermediary). The intermediary node has the key to open this message and route it the rest of the way to the recipient.

5.3.4 *A new layer in the Lightning Network: Channel factories*

Channel factories provide a further innovation to improve the flexibility of payment channels (Burchert *et al.*, 2018). The so-called channel factories refer to a new layer (a factory to generate new channels) that sits between the underlying blockchain layer and the payment channel layer. The aim of the channel factory layer is to enable trustless off-chain channel funding (since opening and closing channels is an on-chain event, which means transaction fees and non-immediate processing).

The channel factory concept involves having a multi-party channel in which new channels are created through off-chain contracts. This contrasts with payment channels, in which payments between channel members are created through off-chain contracts. The channel factory system allows rapid changes to the allocation of funds to channels and reduces the cost of opening new channels. Instead of one blockchain transaction per channel, each user only needs one transaction to enter a group of nodes, and then within the group, the user can create arbitrarily many channels. For a group of 20 users with 100 intra-group channels, the cost of the blockchain transactions might be reduced by 90% as compared with 100 regular micropayment channels opened on the Bitcoin blockchain (Burchert *et al.*, 2018).

The potential benefit is that users can manage the funds that are committed to payment channels with greater flexibility since payment channels lock-up funds for the duration of the channel. For example, if Alice has $5, she can be in a payment channel with Bob, but cannot devote her capital to other projects. In a multi-party channel, Alice has $5 committed to the channel which can be allocated to different projects with greater flexibility. In this sense, channel factories provide a solution to capital allocation in a cash-based economy.

5.3.5 *Smart routing through atomic multi-path routing*

Other blockchain functionality is also being deployed in smart routing solutions, notably atomic swap technology in atomic multi-path routing.

5.3.5.1 *Atomic swaps*

An atomic swap is a smart contract technology that enables the exchange of one cryptocurrency for another without using centralized intermediaries, such as exchanges. The idea is a peer-to-peer exchange of funds of the same value between two cryptocurrencies using a smart contract as an escrow service. A programmatic feature called hash time-locked contracts is used to ensure that both parties fulfill the requirements of the trade. If both parties do not deposit their funds within a certain time parameter, the trade is canceled. In this context, atomic means "all or nothing" or undividable (as atoms are the smallest indivisible material). The two sides of the transaction are necessary to compose the atom (the whole) of the transaction. This prevents one party from tendering their funds and the other party stealing them without fulfilling their part of the deal. The swap is not executed unless both parties deposit their funds.

The notion of atomic transactions is derived from computer science more generally and database management, in which atomic transactions are those that either occur completely or not at all (there is no partial file transfer). Likewise, in blockchains, atomic transfers either happen completely or not at all. Atomic swaps on blockchains were proposed in 2013 and one of the biggest execution venues is the Komodo platform (a multi-chain platform) (Nolan, 2013). Similar kinds of order structures exist in stock market trading such as the all-or-none (AON) directive to fill the order completely or not at all, and a related idea, fill-or-kill (FOK), fill the order immediately or kill it.

5.3.5.2 *Atomic multi-path routing*

The atomic principle in blockchains is being applied to another envisioned upgrade to payment channels, in the form of atomic multi-path routing. In this smart routing concept, instead of routing payment channel transactions through channels that only have the capacity for the full amount, atomic multi-path routing is used. Atomic multi-path routing accommodates the use of multiple channels with smaller amounts as long as the full amount can be delivered with the same atomicity guarantees that all or none settlement provides. In atomic multi-path routing, either all of the smaller channel payments go through and a receipt is produced, or none of them go through and the transaction is canceled and must be resubmitted using another set of routes.

5.4 World Economic History on Replay

In some sense, the migration to digital economic networks can be seen as a replay of the entirety of world economic history in the space of a few years. Traditional solutions are being reinvented for digital networks and innovative new solutions are appearing as well, including EconTech and GovTech applications (the outsourcing of institutional functions to digital networks). The pervasive shift to blockchains can be seen in the United Nations advocating for privacy as a basic human right, and supporting secure multi-party computation and zero-knowledge proofs (UN, 2019). Some of the aspects of traditional financial systems with parallels in blockchain economic networks are listed in Table 5.2.

Initially in blockchains, tokenized forms of money were established, and began to be exchanged between users in the forum of marketplaces. Mechanisms were needed for fundraising, the marshalling of capital, the issuance of money supplies, and investor participation. Interstate, cross-border commerce, and licensing became an issue as existing regulations (related to being a money transmitter) were applied to the new domain.

Table 5.2. Economic themes with instantiations in blockchain networks.

Tokens	Lending	Economic statistics
Exchanges	Regulation	Inflation/deflation
Marketplaces	Licensing	Targeted stimulation
Interstate/cross-border commerce	Speculative bubbles and crashes (cryptocrashes)	Asset-backed lending (USD/ Bitcoin)
Capital-raising	Fractional reserves	Bank runs
Currency minting	Digital assets	Banking the unbanked
Investor participation	Taxation	Arbitrage
Volatility (stablecoins)	Accounting	Contracts & swaps
Global debt registry	Futures and options	Governance & insurance

Securities regulation laws were applied to initial coin issuance and exchanges. Futures, options, indices, and other traditional financial investment products were developed in the blockchain context. Taxation (capital gains and otherwise) and accounting treatment were defined in different countries, establishing whether cryptocurrencies are a property or a currency. Reporting and record-keeping mechanisms were needed. Lending and money-lenders arose, addressing reserve requirements, asset-backed lending, fractional-reserve banking, cross-market lending (USD/Bitcoin), and the need for a global debt registry. Staked coins (asset-backed coins) and staking-as-a-service arose as flexible financial services products.

Speculative bubbles, cryptocrashes, and arbitrage opportunities have characterized the domain. Various classes of digital assets (fungible and non-fungible assets) and fractional asset ownership schemes have arisen. Banking solutions have been made available to a new tier of customers that were unprofitable to serve with the brick-and-mortar bank branch cost structure. Other markets have been served such as legalized cannabis retailers who were unable to obtain traditional bank accounts. Issues such as governance, voting, participation, and credit-assignment have become important attributes of the digital financial system. Innovations in organizational vehicles have arisen such as DAOs (decentralized autonomous organizations). The monetary system became a venue for political expression including opting out of traditional fiat currencies. In one example, there was the idea for end users to wield their market power in the form of a bank run against centralized exchanges. The so-called proof-of-keys movement encouraged users to withdraw their funds and close their accounts on Satoshi day (the January launch anniversary of Bitcoin). Price volatility in cryptocurrencies has been addressed by a new idea, stablecoins.

5.5 Verifiable Markets, Marketplaces, Gaming, Stablecoins

Some of the most innovative models that are emerging in blockchain economics are in the areas of digital marketplaces, verifiable markets, blockchain video gaming, and stablecoins.

5.5.1 *Verifiable markets*

Verifiable markets are trustless markets (markets in which it is not necessary to trust any of the participants) because trust is conferred by proof technology. Whereas blockchains are trustless due to the consensus algorithm process that confirms transactions, verifiable markets are trustless due to verifiable proofs. Verifiable markets use cryptography and game theory to verify the market such that trusting participants is not required. A verifiable market is a new kind of market in which market participants (goods and services providers and customers) are not trusted, only the computational proofs they provide.

A prototype of verifiable transactions is smart contracts. Smart contracts can be used to solve the Amazon problem. The Amazon problem is that it is necessary to trust Amazon or another centralized intermediary in order for remote buyers and sellers to exchange goods. A trustless market with decentralized intermediaries might be preferable. In the current model, Amazon sits as a centralized escrow service between the buyer and the seller. Instead, with blockchains, an Amazon-type transaction between a remote buyer and a seller can be instantiated with smart contracts. The buyer orders the item and submits the payment to a smart contract address that serves as an escrow. The seller ships the item, and upon confirmation of receipt of the item, the escrow contract transmits the payment to the seller. A third-party oracle such as FedEx or UPS online shipment tracking can be used to confirm the item delivery.

Verifiable markets envision a more robust version of market activity between remote participants by using cryptographic proofs to confirm each step of the transaction. The buyer deposits the money with a decentralized third party (which could still be a smart contract), that generates a cryptographic proof (with SNARKs or other proof protocols) that the funds have been deposited. Likewise, a cryptographic proof is generated when the goods have been shipped and delivered.

The implication of using cryptographic proofs as the mechanism for verifiable markets is that new kinds of markets might be enabled. There are many possibilities when considering not only goods but services. For example, in situations of lending practices and government contract bidding, it would be useful to be able to demonstrably prove certain aspects of the service. A lender could generate a proof that certain kinds of services were offered to customers (without disclosing the specific details per zero-knowledge proof technology). Governments could prove adherence to open-bidding practices in awarding contracts.

5.5.2 *Digital marketplaces*

Beyond digital cryptocurrencies, there are other forms of cryptographic assets. One such cryptographic asset is digital collectibles, which are unique digital assets (also called non-fungible tokens) that users collect and trade. One of the first digital collectibles projects is CryptoKitties, digital collectible cats that can be owned and bred (each Kitty comes with its own Ethereum smart contract that stores its unique "catribute" DNA). Breeding new cats expresses different catributes in the offspring. The CryptoKitties project is notable as the first mainstream application of blockchains. CryptoKitties are traded through many different websites and apps, and also digital marketplaces.

The biggest digital marketplace, conceptually analogous to the "eBay of digital collectibles", is OpenSea.io. As of June 30, 2019, OpenSea's website statistics indicated 3,500,000 digital assets for sale in 135 categories, with a completed transaction volume of 20,000 ETH since inception in January 2018 (conservatively, ~US$4 million (20,000 ETH × $200)) (OpenSea, 2019). Various digital collectibles and video game assets are available in the marketplace.

5.5.2.1 *Blockchain video gaming*

Blockchain video gaming refers to video games that are interconnected and facilitated by blockchains. The idea is that players can move assets between worlds and trade with other players directly, a functionality not currently offered by conventional video game platforms. Instead of being locked into one game world, assets acquired in one game could be traded for assets in other games, or even used in other games. The traditional gaming model is that users "borrow" assets from game publishers for their length of use of the platform. In blockchain gaming, however, assets could be tokenized items that are permanent objects owned by users, and are portable across brands, games, accounts, and marketplaces.

Blockchain functionality allows gaming economies to become more complex. Gaming wallets or decentralized applications (DApps) could be used to coordinate new ways of interacting with a game world's ecosystem. The immediate features for such DApps are coordinating player-owned assets, trading items between players in secondary marketplaces, and the interoperability of items between games. In a potential evolution to player/user-owned game economies, one implication is the decentralization of video games, including assets, rewards, narratives, quests, and source code. Community participation is suggested through participatory voting on game modifications and expansion plans. In full-blown decentralized video games, all decisions could be community-determined.

An early blockchain gaming project is Neon District, a gaming engine running on the Ethereum blockchain (Hudzilin, 2019). The engine enables online multiplayer role-playing games and procedural narrative game design. Neon District provides optional economic services for the platform, including digital asset custody. One game built on the Neon District platform is Plasma Bears (similar to a gaming version of CryptoKitties), a collectible crafting and questing game in which users collect bear parts, craft an army, go on quests, and trade bears and parts.

Some of the side benefits of blockchain gaming economies are that they could pave the way for greater consumer adoption of blockchain solutions. Users who feel comfortable with blockchain applications in video game economies might be interested in using the same kind of functionality in other contexts. Another benefit relates to the trend of user-generated content. Leading game engine company Unity finds that their platform is being used for many real-life 3D prototyping applications (for example, in autonomous driving, particle physics, and architectural building design). There is a similar demand to export user-owned assets created in gaming environments for use in other domains such as machine learning and data modeling.

5.5.3 *Stablecoins*

Stablecoins are cryptographic coins that are pegged to another currency (such as USD or Ether) in a stable ratio (usually 1:1) to avoid the price volatility of the underlying cryptocurrency and to make the stablecoin reliable as a medium of exchange. There are centralized and decentralized stablecoins. Centralized stablecoins are those such as JPM Coin that is pegged to the USD. Reserves are kept in USD-denominated bank accounts and the coin is custodian-managed, meaning subject to the full traditional regulatory compliance and oversight of the bank. Decentralized stablecoins are those that maintain the reserves in the underlying cryptocurrency, and have other blockchain-related features. A prominent example of a decentralized stablecoin is the MakerDAO in the Ethereum ecosystem.

The MakerDAO's DAI is a decentralized stablecoin. DAO refers to the kind of organizational structure involved, a distributed autonomous organization (DAO). DAI is the ticker symbol of the MakerDAO's stablecoin. The stablecoin's ticker is DAI as Ether's is ETH. The concept of the DAI stablecoin is to have a greater-than 100% reserve-backed cryptocurrency that is managed automatically by a smart contract.

Interested buyers enter the DAI stablecoin smart contract. In the smart contract, the buyer exchanges Ether for a collateralized debt position (CDP) (MakerDAO, 2017). The buyer chooses a collateralization ratio based on the investment size and receives stablecoin (DAI) in return.

The smart contract issues the DAI stablecoin against the reserve for a certain value. DAI is a stablecoin that can be used as any other cryptocurrency.

The DAI white paper (MakerDAO, 2017) describes the specifics. An example is given for investing 150 Ether into the DAI smart contract. Based on the prevailing USD/ETH exchange rate, the smart contract might issue 100 DAI stablecoin (meaning the stablecoin is collateralized at 150%). There is some room for price volatility. As the price of Ether fluctuates, the difference can be absorbed by the reserves within a certain range. However, if the reserve drops below a certain percentage, the collateralized debt position could be liquidated and lost. Investors must re-collateralize the position with additional capital, or sell back some DAI (taking them out of circulation) in order to maintain the stability of the DAI. Each owner manages this rebalancing activity within their own individual collateralized debt position, and this is the mechanism for maintaining the overall stability of the DAI stablecoin.

The token of the MakerDAO (ticker: MKR) is used to vote on governance issues and the interest rates required to maintain the peg between DAI and USD. The goal is to have a 1:1 peg (one DAI = one USD), backed by Ether in a decentralized smart contract. In actual trading results, DAI has sometimes failed to maintain the peg. At one point, a low of 95–96% parity of the value was reached, but was rebalanced. In general, the ratio is slightly on the downside, with DAI worth less than USD (98% as of June 30, 2019 (CoinMarketCap, 2019)).

Overall, the DAI stablecoin has been successful in general at maintaining stability since its launch in December 2017, reaching the stated goal of a stablecoin, precisely in the face of considerable volatility in the underlying Ether during the same period. One risk is concentration and overuse while the technology is still in the early phases of development and potential software bugs and other problems have not yet been revealed. At some point, over 5% of the total Ether money supply in circulation was locked in DAI. The idea would be to proceed more slowly so as to avoid previous problems like the 2016 DAO hack in which software bugs in newly launched software allowed hackers to steal funds from accounts.

Centralized stablecoins are bank-based digital coins (JPM Coin) and bank-like fintech digital coins (Facebook's GlobalCoin). Facebook's

GlobalCoin (announced to launch in 2020) is a centralized stablecoin, meaning a custodian-run digital money, not truly an open cryptocurrency. GlobalCoins could potentially turn Facebook into a bank, with a 2.38 billion worldwide installed base as of March 2019 (Noyes, 2019) (over 25% of the world's population), and an apparatus for local-law compliance already in place. The success of social network conglomerates providing consumers financial services has been demonstrated in China, where Accenture estimates that digital payment solutions (including apps such as AliPay, WeChat Pay, Tencent, and PayPal) may reduce banking revenues by one third by 2020 (Li, 2015). Although enterprise blockchain adoption has been proceeding quickly, consumer-based blockchain services have not, and social networking platforms could add the user experience layer and context necessary to support consumer blockchain adoption. Blockchain functionality beyond payments could be used to deliver a new tier of aspirational value-added services to social network users, possibly helping to repair the reputational status of these companies (Swan, 2018).

5.6 Consensus

5.6.1 *Next-generation classical consensus*

Consensus refers to the process by which a blockchain network reaches agreement about updates to the ledger, using a software algorithm, and without the involvement of any human parties. The consensus algorithm is the source of the blockchain's trustless trust. The principle behind trustless trust is not to avoid trust, but rather to transform trust from instead of trusting one entity or the parties involved in transactions to trusting a mathematical application, and over time, offloading more and more trust functionality to such verifiable computation mechanisms. Thus far, consensus algorithms have succeeded at providing the high level of cryptographic security that is necessary for digital monetary systems, but at an expensive cost. Bitcoin's over 10-year history demonstrates the success of digital consensus-based value transfer (Bitcoin's first transactions were in January 2009, and the chain has logged over 583,000 blocks of transactions as of June 30, 2019 (Blockexplorer, 2019)). The Bitcoin blockchain has not been successfully attacked (the attacks that have happened are typically at the blockchain on-ramps and off-ramps of user wallets and exchange websites). The downside is that the cost of providing cryptographic security through the competitive proof-of-work mining effort is expensive and perhaps not scalable, and therefore next-generation consensus algorithms seek to provide the same level of cryptographic security with greater efficiency.

There are a number of projects underway for next-generation consensus algorithms, including DFINITY, Hashgraph, IOTA, Nano, and ByteBall. Such algorithms would purportedly improve the scalability of existing blockchains. Many next-generation algorithms are in the form of Practical Byzantine Fault Tolerance (PBFT), meaning that a network of distributed computers can reach a consensus about system updates even in the face of adversaries. Like many issues in the blockchain context, the problems are known challenges in a variety of computer science, cryptography, and network science fields. Of specific focus are computer science problems such as fast Byzantine agreement and the leader election problem. Classical and quantum solutions have been proposed.

5.6.2 *Next-generation PBFT: Algorand and DFINITY*

Two notable next-generation consensus projects with available solutions are Algorand and DFINITY. Both rely on randomness generation and proof verification. The law of large numbers is used, in that from a large group of nodes, a certain number are selected at random to confirm each block. Algorand is an open solution, whereas DFINITY is proprietary. These next-generation methods could ideally provide secure cost-efficient consensus to blockchains with millions of users.

Algorand (algorithm providing randomness) uses a fast Byzantine agreement protocol as the consensus algorithm. Unlike Bitcoin, the agreement is not performed between all of the users in the network, but rather confined to a small committee of randomly chosen users for each round. A decentralized voting mechanism is used to pool and randomly select users to develop a unique committee to approve each block. The approach is based on Verifiable Random Functions (Micali *et al.*, 1999). A verifiable random function is a cryptographic building block that maps inputs to verifiable pseudorandom outputs. Verifiable random functions extend earlier work in cryptography known as the Goldreich–Goldwasser–Micali construction of pseudorandom functions and zero-knowledge proofs (Goldreich *et al.*, 1986). The process combines unpredictability and verifiability. Algorand uses verifiable random functions to perform cryptographic sortition to select committees to run the consensus protocol (this is called the leader selection problem in computer science). Sortition means selecting a random sample from a larger group. The decentralized voting mechanism pools and randomly selects available users to develop a unique committee to approve each block.

DFINITY also relies on randomness, by including a scalable decentralized random beacon directly in the consensus protocol. The project uses a threshold relay technique for decentralized, deterministic randomness which is made possible by certain characteristics of the Boneh–Lynn–Shacham (BLS) signature system used in the algorithm. The signature scheme uses a bilinear pairing for verification in which signatures are elements of an elliptic curve group. The DFINITY consensus algorithm also uses a notarization technique that overcomes some of the traditional problems in proof-of-stake systems. Security proofs are also incorporated (Hanke *et al.*, 2018). The general concept is that 250 of 300 signatures are required to confirm a block, and that the 300 possible signers have not participated in a recent consensus process.

5.6.3 *Quantum Byzantine Agreement*

Another line of consensus algorithm development specifically targets post-quantum security. Fast Byzantine Agreement is an ongoing area of computer science research. (Byzantine Agreement refers to distributed computers in a network reaching agreement about updates irrespective of adversaries.) Fast Byzantine Agreement uses faster time processing algorithms than regular Byzantine Agreement. Such faster time processing algorithms may be polylogarithmic as opposed to logarithmic (the Hashcash proof-of-work algorithm used by Bitcoin is logarithmic). Other complexity features are proposed, for example, a probabilistic Byzantine Agreement algorithm in which both the time and the communication algorithms are polylogarithmic (Braud-Santoni *et al.*, 2013). The algorithm is based on an almost everywhere-to-everywhere agreement protocol (by analogy to a complete graph in network theory).

Various methods for fast quantum Byzantine Agreement algorithms are also proposed. One such fast quantum Byzantine Agreement algorithm claims to offer a substantial speed-up over classical methods, including in situations of dynamic adversaries and faulty nodes (Ben-Or & Hassidim, 2005). Other work addresses the topic of distributed consensus algorithms on quantum networks. Whereas most of the progress has been focused on optimizing the convergence rate of the algorithm for quantum networks with undirected underlying topology, this approach uses directed underlying graphs (Jafarizadeh, 2017).

References

1ML (2019). Lightning Network Search and Analysis Engine. https://1ml.com/. Accessed June 30, 2019.

Ben-Or, M. & Hassidim, A. (2005). Fast quantum byzantine agreement. In: *STOC '05 Proceedings of the Thirty-seventh Annual ACM Symposium on Theory of Computing*. Baltimore, MD, USA, May 22–24, pp. 481–485.

Blockexplorer (2019). https://blockexplorer.com/. Accessed June 30, 2019.

Braud-Santoni, N., Guerraoui, R. & Huc, F. (2013). Fast byzantine agreement. In: *PODC '13 Proceedings of the 2013 ACM Symposium on Principles of Distributed Computing*. Montréal, Québec, Canada, July 22–24, pp. 57–64.

Burchert, C., Decker, C. & Wattenhofer, R. (2018). Scalable funding of Bitcoin micropayment channel networks. *R. Soc. Open. Sci.* 5(8):180089.

CoinMarketCap (2019). https://coinmarketcap.com/currencies/dai/. Accessed June 30, 2019.

Danezis, G. & Goldberg, I. (2009). Sphinx: A compact and provably secure mix format. In: *30th IEEE Symposium on Security and Privacy*. May 17–20, pp. 1–14.

Goldreich, O., Goldwasser, S. & Micali, S. (1986). How to construct random functions. *JACM* 33(4):792–807.

Hanke, T., Movahedi, M. & Williams, D. (2018). DFINITY Technology Overview Series: Consensus System. Rev.1. Technology Overviews: DFINITY Stiftung.

Hudzilin, A. (2019). The State of the Blockchain Gaming Industry. Medium.

Jafarizadeh, S. (2017). Optimizing the convergence rate of the continuous-time quantum consensus. *IEEE Trans. Autom. Control*. 12:6122–35.

Li, L. (2015). Are WePay and Alipay going to kill banks? Walk the Chat.

MakerDAO (2017). The Dai Stablecoin System. https://makerdao.com/whitepaper/DaiDec17WP.pdf. Accessed June 30, 2019.

Micali, S., Vadhan, S. & Rabin, M. (1999). Verifiable random functions. In: *FOCS '99 Proceedings of the 40th Annual Symposium on Foundations of Computer Science*. October 17–18, pp. 120–31.

Nolan, T. (2013). Alt chains and atomic transfers. Bitcoin Forum.

Noyes, D. (2019). The Top 20 Valuable Facebook Statistics. Zephoria.

OpenSea (2019). https://opensea.io/. Accessed June 30, 2019.

Poon, J. & Dryja, T. (2016). Lightning Network paper, v0.5.9.1. https://cryptochainuni.com/wp-content/uploads/Bitcoin-lightning-network-paper-DRAFT-0.5.pdf. Accessed February 18, 2020.

Swan, M. (2018). Blockchain consumer apps: Next-generation social networks (aka strategic advice for Facebook). CryptoInsider.

United Nations (2019). UN Handbook on Privacy-Preserving Computation Techniques. Big Data UN Global Working Group.

Valloppillil, V. & Ross, K. (1998). Cache Array Routing Protocol v1.0. Expired Internet Draft. https://tools.ietf.org/html/draft-vinod-carp-v1-03. Accessed June 30, 2019.

Chapter 6

Quantum Blockchain

Abstract

The future of global network communications could include a quantum internet with various features such as quantum key distribution (QKD), secure end-to-end communication, quantum memories, quantum repeaters, and quantum-based applications such as quantum blockchains. Reactions toward the potential quantum information era are first, one of preparing to be quantum-resistant and quantum-secure, and second, being quantum-compatible and quantum-embracing by taking advantage of the new functionality offered by quantum systems. There are implementation suggestions for quantum blockchains, both the overall protocols and individual aspects such as cryptography upgrades. The nearest-term potential application of quantum computing appears to be QKD as an antidote to RSA cryptographic standards possibly being compromised by quantum computers. Specific ways in which blockchains can become more quantum-secure are considered.

6.1 Quantum Blockchain

Quantum blockchain refers to the idea of either an entire blockchain or certain elements of the blockchain functionality being instantiated and run in quantum computing environments. In fact, the quantum domain naturally lends itself to the implementation of blockchain features. This is through quantum key distribution (QKD) and quantum signatures (post-quantum cryptography), certifiable randomness and fast Byzantine Agreement (scalable consensus), built-in zero-knowledge proof technology (through the QSZK (quantum statistical zero knowledge) computational complexity class), the no-cloning theorem (cannot copy (i.e. double-spend) assets) and the no-measurement rule (cannot look at quantum information or eavesdropping is evident). Since the quantum domain is conducive to the functionality needed by blockchains, and more importantly, because blockchains must articulate a quantum-secure upgrade path, a number of early-stage quantum solutions have been proposed. As in migration to the quantum domain more generally, the first step is to replace the blockchain features that are known to be at quantum risk, which are the cryptographic algorithms used by many blockchains. In the longer-term, next-generation projects might exploit the advantages of the quantum domain by instantiating full blockchain protocols with quantum information theoretic principles.

6.1.1 *Quantum-secure blockchains and quantum-based logic*

The first and most basic idea for quantum-secure blockchains is upgrading the cryptography to QKD (producing and distributing keys generated by quantum computers). At least one method for quantum-secure blockchains using QKD, in an urban fiber optic network, has been proposed (Kiktenko *et al.*, 2018). Other research builds on this to articulate a more extensive quantum logic-based blockchain with robust features and a native token called Qulogicoin (Sun *et al.*, 2019). Blockchain protocols would be translated into a framework of quantum circuits. One quantum circuit could implement the consensus algorithm, replacing the classical Byzantine Agreement protocol with a quantum Byzantine Agreement protocol. Another quantum circuit could encode quantum certificates and other quantum protection methods into the transaction syntax.

Another sophisticated idea for implementing blockchains with quantum-based logic relies upon entanglement. The project envisions a temporal Greenberger–Horne–Zeilinger (GHZ) blockchain in which the functionality of time-stamped blocks and the hash functions linking them is replaced with a temporal GHZ state which is entangled in time. The quantum system chains the bit strings of the Bell states together in chronological order, by being entangled in time. The temporal Bell states are recursively projected into a growing GHZ state. As in other situations of entangled quantum information, an attacker attempting to tamper with the photons would be immediately detectable and invalidate the state (Rajan *et al.*, 2019). Secure quantum protocols are thereby provided.

6.1.2 *Proposal for quantum Bitcoin*

Another proposal for a full quantum blockchain solution is qBitcoin (quantum Bitcoin), "a peer-to-peer quantum cash system" (Ikeda, 2017), making reference to the initial Bitcoin white paper for "a peer-to-peer cash system" (Nakamoto, 2008). The project has a number of quantum-based features. Also parallel to the original Bitcoin white paper, the paramount concern is preventing double-spending. In Bitcoin (and other blockchains), double-spending is prevented by a global timestamping system (Bitcoin is essentially a clock), and an always-on worldwide network (facilitated by the internet) that checks in real-time whether a unique currency balance is available whenever there is an attempt to spend it. In qBitcoin, double-spending is envisioned to be prevented by using quantum teleportation technology for transactions (the transmission of an exact state of quantum information), which would prevent the owner from keeping coins once they are spent (after the quantum state is sent). Other quantum features used in the qBitcoin system are quantum digital signatures and a data structure based on a quantum chain instead of conventional blockchain blocks, which are time-consuming to assemble.

6.1.3 *Quantum consensus: Grover's algorithm, quantum annealing, light*

Since classical consensus (i.e. proof-of-work mining) is not the most scalable and efficient of systems, there are many ideas for implementing consensus in quantum formats. One suggestion is to speed up the mining process by using a modified Grover's algorithm (used in large data searches) (Sapaev *et al.*, 2018). A procedure is elaborated, starting with transforming the nonce register into uniform superposition states, and applying a series of steps to implement a Grover's diffusion operator (one of the two important gates in Grover's algorithm).

Another idea proposes using quantum annealing machines for quantum consensus, implementing the proof-of-work algorithm with Hamiltonian optimizers (Kalinin & Berloff, 2018). A third proposal sets forth an idea for a quantum-enabled blockchain using light, and describes more generally how quantum optical devices (such as quantum modems) might be integrated into blockchain architectures (Bennet & Daryanoosh, 2019). The consensus mechanism would be based on proof-of-entanglement. Clients and servers would participate in an interactive mining protocol (a three-round process of authentication, mining, and consensus) to generate and commit entanglement as a resource towards securing the blockchain. The proof-of-entanglement algorithm is developed as a variation of the Einstein–Podolsky–Rosen (EPR) steering protocol (a means of nonlocally affecting another state in a quantum system through local measurements).

6.1.4 *Quantum money*

Quantum money has been proposed for some time, based on the no-cloning theorem, which prevents quantum information from being copied (Wiesner, 1983). Quantum information that cannot be copied suggests that double-spending is not possible and that assets are unique, crucial features in a monetary system. However, both the initial idea and most other proposals rely on a central bank to do the currency issuance and verification. This includes a recent suggestion for the preparation and verification of unforgeable quantum banknotes (Guan *et al.*, 2018). A key property of blockchains is decentralization. Hence, other work indicates that it would be possible to have a publicly verifiable (by any party) quantum money based either on quantum oracles or on random stabilizer states (Aaronson *et al.*, 2012). The mechanism might be further instantiated for quantum states to be used as copy-protected programs.

6.2 Quantum Internet

The quantum internet is the concept for a future internet based on quantum technologies. The quantum internet would be an extremely secure network that operates by exploiting the effects of quantum mechanics. Worldwide quantum networks would be more secure and also much faster than today's internet. Some of the first envisioned applications of the quantum internet are QKD and secure end-to-end communication. The initial internet was not designed to be secure, and hence, the quantum internet is a privacy, security, and efficiency upgrade for the existing infrastructure. Quantum smart network is the further extension of the quantum network concept in the idea of smart network technologies (such as blockchain and deep learning) running on the quantum internet. The quantum internet could render the internet itself as a smart network by virtue of having a richer suite of automatically executing computational protocols built directly into its operation. The potential progression is one in which communications networks become computation networks, and then become quantum computation networks. A roadmap for the potential development of the quantum internet has been proposed as highlighted in Table 6.1 (Wehner *et al.*, 2018, p. 4).

Table 6.1. Roadmap: Six steps to a quantum internet.

Step	Functionality	Description
1.	Trusted-node network	End users can receive quantum-generated codes, but cannot send or receive quantum states; any two end users can share an encryption key.
2.	Prepare and measure	End users can receive and measure quantum states (without involving entanglement); end user passwords are confidentially verified.
3.	Entanglement distribution networks	Any two end users can obtain entangled states (but not store them), these providing the strongest quantum encryption possible.
4.	Quantum memory networks	Any two end users can obtain and store entangled qubits (the quantum unit of information), and can send quantum information to each other; network-enabled cloud-based quantum computing.
5.	Quantum computing networks	The devices on the network are full-fledged quantum computers (able to process error correction on data transfers); distributed quantum computing and quantum sensors enable applications for science experiments.

6.2.1 *Quantum network theory*

Quantum network theory is a theory of quantum network computing that describes a standard set of protocols, components, and processes for the realization of quantum networks. Such standard components of quantum network infrastructure may include quantum processors, quantum repeaters, quantum memory, entanglement generators, and cloud-based quantum computing services. Various areas of quantum information science (such as quantum computing, quantum networks, and quantum cryptography) are defining relevant applications. Such applications include using quantum coding theory to develop quantum stabilizer codes for use in secure end-to-end communication, and quantum error correcting codes for use in quantum computing (Wilde, 2008).

6.3 Quantum Networks: A Deeper Dive

The concept of quantum networks started to be discussed in 2008 (Kimble, 2008). It was proposed that a quantum network must be able to generate, distribute, and process quantum information in addition to classical information. Such quantum networks would need to generate and transfer quantum coherence and entanglement, and convert quantum states from one physical system to another, possibly through the optical interactions of single photons and atoms.

6.3.1 *The internet's new infrastructure: Entanglement routing*

The quantum internet is conceived as an extension to existing internet infrastructure. An important feature upgrade is entanglement routing, which is the ability to route (transfer) quantum entanglement across the internet. Such routed entanglement on the quantum internet could be used in various applications, most notably secure end-to-end communication between two parties. A new line of quantum internet equipments such as quantum repeaters and quantum modems might be needed to implement this kind of model of entanglement routing for widespread communication.

Entanglement is a unique feature of quantum mechanics which allows particles with two distinct quantum states to have a closer relationship than is possible in classical physics. A key point is that if two particles are entangled, then the state of one particle can be somewhat known by measuring the state of the other. This is useful for building fault tolerant quantum information systems since quantum information states can be error-corrected through the entanglement property. Entanglement is also implicated in quantum security since the very act of measuring the state of a quantum system disturbs it. Entanglement can be used to indirectly check the states of particles, and to detect the presence of a third-party eavesdropper. Entanglement as a feature of quantum internet communication is thought to be able to deliver virtually unbreakable privacy.

Routing protocols in quantum networks are active research frontiers. Various quantum entanglement routing protocols are proposed for generating entanglement between multiple pairs of users in a quantum network. In such quantum networks, each repeater node would be equipped with quantum memories and entanglement sources. Whereas linear routing is a current standard in optical networks, nonlinear routing might offer better scaling for quantum networks. One research project suggests that nonlinear multi-path routing might offer superior performance over single-path routing (Pant *et al.*, 2019). Using multiple paths for routing entanglement between a pair of end users might enable long-distance entanglement generation with better scaling than is possible with only a single linear repeater chain (which routes along the shortest path between the users). This is a counterintuitive finding because linearity is a typical design property in optical networks (light travels linearly). The multi-path routing result is exemplar in engaging ideas from multiple fields to study the new domain of quantum networks. Some of the other ways the quantum network design space is being expanded is by including other canonical ideas from network science such as any-to-any complete graph connectivity and directed graphs as routing protocols.

Multi-path routing invokes the notion of superposition, in that the routing could take place over any number of configurations or paths through the network, similar to the way a particle could be in any state within the Hilbert space. Just as input data and logic gates might be superpositioned in quantum photonic networks (Procopio *et al.*, 2015), the same could be true for routing in the quantum internet. Superpositioning in multi-path routing could be incorporated into quantum algorithms for improved routing performance in the context of quantum smart routing. Quantum smart routing is the notion of smart routing on quantum networks based on superposition.

6.3.2 *Quantum memory*

In photonic quantum computing, a quantum memory is an interface between light and matter that allows for the storage and retrieval of photonic quantum information, analogous to the memory in a conventional computer. Somewhat similar in principle to quantum error correction, a quantum state of light is mapped onto an ensemble of atoms and then recovered in its original shape. For the sophisticated applications of long-distance quantum communication, a quantum memory system is likely to be necessary (Le Gouet & Moiseev, 2012). Recovering a stored quantum state of light is non-trivial. The idea is that the quantum internet would allow the secure exchange of information represented by quantum superposition states stored in quantum memories. Quantum memories with short coherence times have been demonstrated with matter-based quantum computing techniques using atoms and ion traps.

A prominent form of quantum memory is topological quantum memory (Dennis *et al.*, 2002). A topological quantum memory is again in the form of a quantum error-correcting code. Qubits are arranged on a topological surface (such as a Majorana chain) and encoded for quantum operations. As the error rate gets higher and reaches a critical value, a phase transition occurs in the system between the ordered state and a disordered state. The phase transition can be modeled as a lattice gauge theory with quenched disorder. As long as the error rate stays below the critical value, the encoded information can be protected, or used as a quantum memory. Other research suggests that the ability to perform active quantum error correction on topological codes might also be a good possibility for long-term qubit storage (Lang & Buchler, 2018).

6.3.2.1 *All-photonic quantum optical repeaters*

The internet comprises a global network of fiber optic cables. Light loses intensity as it travels long distances. Repeaters are needed to boost and amplify the optical signal, and are inserted at regular intervals along optical cable transmission lines. The quantum version of the internet might replace the existing optical repeaters with a new kind of technology, all-photonic quantum repeaters. All-photonic quantum repeaters use components that are based only on optical devices such as linear optical elements, single-photon sources, and photon detectors. One benefit of all-optical components would be that the communications system is more fault-tolerant and efficient.

At present, quantum repeaters have been built, but are not all-photonic devices. They comprise a mix of matter-based quantum memories and optical devices. Such conventional quantum repeaters are difficult to implement because they need to store a quantum state at the repeater sites, which is expensive and requires cryogenic temperatures. A recent advance is the experimental demonstration of quantum optical repeaters using an all-photonic protocol (Hasegawa *et al.*, 2019). The research conducted a standard procedure with the all-photonic quantum repeaters known as the time-reversed adaptive Bell measurement.

6.4 Quantum Cryptography and Quantum Key Distribution

Quantum cryptography is an area of computer science, mathematics, and physics in which quantum mechanical properties are exploited to perform cryptographic tasks. There are many different potential applications of quantum cryptography, particularly as related to network communications and smart network technologies. The biggest potential short-term application is QKD, which solves the problem of transporting quantum cryptographic keys securely between remote locations. QKD is purported to be extremely secure because any third-party intrusion is detectable. Other applications include fast Byzantine Agreement, quantum entanglement, certifiably random bits, and quantum statistical distributions. Whereas current cryptography standards rely on math (the difficulty of factoring large prime numbers and other known problems that are difficult to solve) which might be broken with quantum computers, quantum cryptography relies on physics (the quantum states themselves), which are not breakable because they are the most foundational form of physics.

6.4.1 *Quantum key distribution*

Although quantum cryptography includes many applications, the first and foremost is QKD. QKD appears to be one of the most likely quantum technologies to be adopted in the short-term, on top of otherwise existing internet infrastructure. The global communications network uses key distribution as a secure method of encryption to protect everything from credit card transactions to passwords and SMS texts. The further implied use of key distribution is to make every use case of user identity and credential verification more secure. At present, personal information must be revealed for various purposes such as to validate a bank account, health record, or other accounts. With QKD as a built-in feature of the communications infrastructure, possibly used in conjunction with proof technology, validation could occur without having to disgorge personal information. The aim is to make the internet more private and secure.

QKD is trying to solve the problem where classical methods for key distribution are insecure. At present, keys are sent as classical bits in a stream of electrical or optical pulses representing 1s and 0s that can be read and copied. It is physically difficult to detect the presence of an intruder when communicating through a classical communication channel. Also, classical methods of key distribution may be broken if enough computing power is available. QKD ensures that an eavesdropper can succeed only with a very low probability, and also that no amount of computing power will allow a QKD protocol to be broken. Any attempt to intercept the quantum key would collapse the quantum state, destroying the information and signaling the presence of an intruder. This means that quantum distributed keys are far more secure than keys sent as classical bits. QKD uses quantum methods in both steps of the process, key generation and distribution.

Classical keys are physically instantiated as classical bits that are 1s and 0s that can be copied. Quantum keys are physically instantiated as quantum bits (qubits) that are quantum mechanical states of photons or atoms that cannot be copied (per the no-cloning theorem of quantum information), and cannot be measured (because it will damage the qubits). Thus, quantum keys are secure because any tampering would be immediately known. If this were to happen, the users would just restart the process and generate another quantum key for their communication. Once the quantum key is established between the sender and the receiver, it is translated into classical bits for the usual encryption and decryption of messages sent across the open internet to continue as per the traditional method. QKD is only used for establishing the private key between the users. Since the key is produced and distributed with quantum methods, it is assumed to be secure for its subsequent use in that session of message sending.

Key exchange refers to the exchange of a private key between two users. The private key is used to encrypt messages that are sent across the open internet, and decrypt messages received from the other party. Typically, a private key is exchanged at the beginning of the communication, used for the duration of the interaction, and then discarded. QKD establishes highly secure keys between distant parties by using single photons to transmit each bit of the key (Jennewein *et al.*, 2000). A private key is generated iteratively between two parties. The key generation process between Alice and Bob could be as follows. Alice sends a single bit (a photon) at a time, measuring certain information parameters about each photon sent. Bob receives and records information about each photon received. At some point, the two parties compare information about what has been sent and received (not the information itself because that cannot be inspected without damaging it, but the metadata about the information that has been sent). When half of the bits are verified, the rest of the stream is dropped and the current bits that have been sent and received are used as a key. The keys are not exactly the same. Alice and Bob have slightly different bits due to noise and the possibly of eavesdropping. If the error rate between the keys is low enough (typically 20%), the key is deemed secure and communication proceeds (Bennett, 1992).

6.4.2 *Satellite-based quantum key distribution: Global space race*

QKD has been demonstrated with many approaches. The next phase could be ramping up to commercialization. Experimental results have been obtained using existing terrestrial communications networks, and in airborne and satellite-based communications. Various teams have demonstrated QKD in existing urban fiber optic communication networks. One Moscow-based project developed a QKD network based on the paradigm of trusted repeaters and a common secret key generated between users via an intermediate trusted node (Kiktenko *et al.*, 2017). The technical mechanism uses polarization encoding and phase encoding. Teams in China and Japan have demonstrated satellite-based QKD with the Micius, Tiangong-2 Space Lab, and SOCRATES satellites (Khan *et al.*, 2018). Teams in Germany and Canada have demonstrated QKD links between airplanes and ground stations with the flying entity serving as both the sender and the receiver of the QKD (Pugh *et al.*, 2017). The feasibility of satellite-based quantum communications is noted in a 2010 paper from Los Alamos National Laboratory (Hughes *et al.*, 2010).

The reason that space-based QKD is desirable is that quantum computers perform better and are less prone to environmental noise and errors in the colder, lower-gravity environment of space. Another benefit is that satellites are already a central fixture in the global communication network, and so integrating satellite-based QKD might proceed quickly and receive widespread adoption. Since the global communication network is already heavily satellite-based, any feature upgrade could roll out nearly immediately on a worldwide basis.

An important satellite-based QKD demonstration was realized in September 2017 with the world's first quantum-encrypted intercontinental video link (Giles, 2018). Quantum keys were distributed from the Chinese satellite Micius to generate a secure videoconference between Vienna and Beijing. The encryption keys (photons encoded in a quantum state) were beamed down by satellite to both participants. The video call was a collaboration between Anton Zeilinger, a quantum physicist at the Austrian Academy of Sciences responsible for the initial Bell's inequality demonstrations (Weihs *et al.*, 1998), and a former graduate student of his, Jian-Wei Pan, a professor at the University of Science and Technology of China in Hefei.

So far, QKD demonstrations have mostly relied upon dedicated hardware, but other projects are using off-the-shelf photonic equipment. Photonic QKD uses the quantum properties of light such as polarization to encode and send the random key needed to decrypt encoded data.

Recent results show QKD based on the polarization of light using commercially available components (Agnesi *et al.*, 2019).

6.4.3 *Key lifecycle management*

The possibility of QKD raises the question of standardized practices more generally regarding key use and lifecycle. Although many keys are for one-time use, other keys are issued for persistent use, but may not be part of a lifecycle management program. For example, blockchains do not typically have any deactivation policy for private keys. The US NIST has published guidelines for defining appropriate crypto-periods for cryptographic keys (Barker *et al.*, 2016, p. 64). The crypto-period is the period of time between the key activation and the key deactivation (i.e. the lifecycle of the key). The NIST report suggests that private signature keys should expire or be deactivated relative to the sensitivity of their use, not necessarily based on a fixed amount of time. Other factors might concern the risk of key compromise and the cost of new key generation. Aside from the crypto-period, there are other issues to consider in an overall key management program. These could include the types of keys (public, private, and symmetric) and their corresponding usage (authentication, authorization, signing, and verification). Further, there are considerations relating to how a key is properly generated, distributed, stored, replaced, deleted, and recovered during its lifetime, and how it is protected against threats.

6.5 Quantum Security: Blockchain Risk of Quantum Attack

On the one hand, blockchains are at greater potential risk from quantum computing than other technologies because they are heavily dependent upon cryptography. The very premise of blockchain protocols is the computational infeasibility of inverting certain one-way hash functions, but these may be broken with quantum computers. On the other hand, blockchains also stand to potentially benefit the most from the innovations developed in quantum cryptography.

Estimates vary as to when quantum computing will be a threat to the current cryptographic infrastructure, blockchain-related and otherwise. An estimate from NIST, drawing from industry experts, suggests that quantum computers powerful enough to break the current 2048-bit RSA standard might be available by 2030 (Chen *et al.*, 2016). Others predict that the elliptic curve signature scheme currently used by Bitcoin (ECDSA) is at even greater risk, and could be broken by quantum computing as early as 2027 (Aggarwal *et al.*, 2018, 1). A planned upgrade to Schnorr signatures may serve to mitigate this somewhat, but is not a robust quantum-resistant solution. To be sustainable in the long term, it is important for blockchains to establish an implementation roadmap for quantum-resistant solutions (solutions that make the system resilient to the kinds of attacks that might be possible with quantum computing).

In general, the kinds of quantum computing algorithms required to break public cryptography and blockchains are not those used in contemporary NISQ devices, but are rather on the order of Shor's algorithm and Grover's algorithm, which are not immanent and will likely require significant numbers of qubits and also quantum error correction. Given the complexity of blockchains, early applications in quantum computing would more likely focus on cryptographic problems that are easier to solve. To the extent the nascent quantum computing industry continues to develop, various general-purpose solutions may arise, and could be implemented in many industries, including blockchains. A prime example of a potential early application of quantum computing is QKD. If such technology were to become the norm for key distribution, blockchains might likely incorporate it into their key issuance procedures in quantum wallets.

Some of the potential risks to public blockchains such as Bitcoin are in the areas of authentication (transactions) and mining (keeping the chain secure). One opinion is that substantial breakthroughs in quantum algorithms would be required in order to reverse existing hash functions (Tessler & Byrnes, 2018). Other research points out that public keys are already vulnerable at present since they are openly exposed during some parts of the normal operations of blockchains. This occurs when coins are spent in proof-of-work blockchains and are used to stake a vote in proof-of-stake blockchains. The work calls for more secure public keys even before any possible quantum information era (Kelly *et al.*, 2018).

The overall situation seems to indicate that the kinds of quantum risks that blockchains face are specific and less damaging than might have been thought, as long as there is some plan for upgrading to post-quantum cryptography, and other upgrades and fixes continue to progress. Blockchains already have some degree of inherent protection built into their structure due to their nature as mechanistic execution networks with numerous restrictions. Ratcheting up the thresholds on some of the in-built protections could be the first line of defense in making blockchains deliberately quantum-resistant. Also, some features that are known to make blockchains vulnerable to quantum attacks at present may already be in the process of being addressed in expected upgrades motivated by other reasons. For example, implementing non-reusable components (addresses, keys, and signatures), private transactions (sender and receiver addresses, and the amount are all masked), and pervasive zero-knowledge proof technology might serve to reduce some of the current quantum vulnerabilities of blockchains. Authentication and mining as two specific risks to blockchains are discussed, along with potential solutions. Both stress the benefit of switching to quantum-resistant cryptographic algorithms.

6.5.1 *Risk of quantum attack in authentication*

Authenticating the user's ownership of cryptocurrency coins or other digital assets at the moment of spending or transfer is an area of blockchain security with potential vulnerability to quantum attack. Cryptographic security algorithms rely on computational problems that are difficult to solve but easy to verify, such as discrete logarithms and factoring. These problems are difficult to solve with classical computing, but may be solvable with quantum computing.

In particular, Bitcoin uses the Elliptic Curve Digital Signature Algorithm (ECDSA) for authentication. ECDSA is based on the classical difficulty of the discrete logarithm problem. The authentication algorithm operates by hashing the public and private key together into an address. At present, the standard is for most wallets to generate a new address with every transaction. In the transfer, cryptocurrency coins are moved from one public–private key pair with a published public key, to a new public–private key pair with a non-published public key. This means that the new private key is secure against potential ECDSA attacks. Hence, existing safeguards mean that private keys in Bitcoin are generally protected against attack. However, despite not being exposed most of the time in Bitcoin, public keys become open when the user spends cryptocurrency coins, and this makes the public–private key pair vulnerable to quantum attack during this time.

The risk of attack is related to Shor's algorithm purportedly being able to efficiently factor large prime numbers on a quantum computer. This poses a risk to existing public–private key infrastructure which is based on the classical difficulty of guessing large prime numbers. The implication is that Shor's algorithm could allow an adversary to reverse the public keys in the mathematical algorithm to determine the corresponding private keys. In the interim period during which a public key is revealed in a pending transaction and before the transaction is added to a block, a quantum adversary may be able to reverse the public key and discover the private key. The private key is vulnerable to quantum attack during the transaction time window. If an adversary were able to recover the private key, a transaction could be submitted from the original wallet with a very high transaction fee. This would provide an incentive to the miners to insert that transaction ahead of the original, resulting in the theft of cryptocurrency from the hacked address. The same kind of exposure risk to public keys likewise occurs in proof-of-stake systems, in which voters publish their public keys alongside their vote in order to validate the vote.

An additional (somewhat smaller) threat is that any transactions made at any point in the past not using automatic key generation might be susceptible to a retroactive quantum attack. For example, if someone executed a transaction a few years ago, and then reused the same address, the balance could be at risk. Even if quantum computers do not exist immediately, to the extent that they can break ECDSA cryptography in the future, attackers could steal cryptocurrency at addresses that have been reused. Since the entire blockchain history is publicly available, an adversary with a powerful enough quantum computer could review the ledger and possibly steal balances from all of the public addresses that have ever been reused. Despite the current standard of using new addresses for each transaction in Bitcoin, this was not true in the past and coins at lots of addresses could be at risk. The impact of this particular threat might be muted, in that already, and certainly by the time quantum computing might be available, many of the balances may have been re-spent through new wallets to new addresses controlled with new keys. Also, such older cryptocurrency coins at reused addresses might be irrecoverable due to other reasons such as key loss (key pairs left in old wallets on old phones).

6.5.2 *Risk of quantum attack in mining*

Following authentication, mining is the other main aspect of blockchains that might be vulnerable to quantum attack. Mining is the consensus process that validates new transactions and keeps the blockchain secure.

One way would be to attack the hashing algorithm by which the mining operation is conducted. In Bitcoin, the hash function, though, is quite strong, and possibly more quantum-resistant than other cryptographic algorithms used in blockchain operations. Bitcoin's Hashcash proof-of-work consensus algorithm uses a double SHA hash function, meaning two sequential applications of SHA-256; a SHA-256 of SHA-256 (a composite function of SHA-256(SHA-256(x))) (Kelly *et al.*, 2018, p. 7; Aggarwal *et al.*, 2018, p. 2). There is no known classical or quantum algorithm that could break it. The double SHA hash function would likely be more difficult to break than many other algorithms that might be more easily exploited with a quantum computer.

Another risk with mining is that miners using quantum computers could launch a 51% attack. A 51% attack is the situation where a single entity controls more than half of the computational power of the blockchain. However, one counterargument is that the computing power behind the Bitcoin network is simply too large for a 51% attack, even with quantum computers. As of June 2019, the hash rate exceeded 50,000 TH/s (tera hashes per second) on the Bitcoin network (Blockchain.com, 2019).

Even with the assumed improvements offered by quantum computing, a malicious party would still need an unrealistic amount of quantum computing power to mount a 51% attack. Further, there are limits to the damage that could be accomplished in a 51% attack. There are many things that are not possible to change such as existing wallet balances and the consensus algorithm. Refusing currently unconfirmed transactions is one of the most disruptive possibilities of a 51% attack. Further, sustaining a 51% attack is also thought to be difficult given the hash rate of the system. Overall, a quantum attack on mining that would be able to undermine the hashing power of the network may not be extremely likely.

Quantum computers might also be used in mining, not to take over the network, but to mine more expediently. Quantum computing could enable more efficient mining by allowing more random nonces to be checked in parallel, for example, using Grover's algorithm. Grover's algorithm takes advantage of superposition states to search through a larger space more quickly. However, the cost of using quantum computers for mining might be too high. There is a trade-off between the potential cost of quantum computers and their potential benefit. Even if available, quantum computers might be too expensive for the economic incentive structure of cryptocurrency mining. Even after quantum computers started to become available, it could take time for them to decrease in cost such that mining would make sense. The expense for a quantum computer might be high enough that the expected overall return from mining on a classical computer would still be higher. The current method of Bitcoin mining with specialized ASIC chips is highly optimized for the task, and classical solutions might have better performance as compared with the estimated improvement of quantum computing. Mining on a quantum computer might be more relevant for other blockchains that do not already have custom solutions (for example, Ethereum, which is memory-hard, i.e. specifically resistant to mining with custom chips).

6.6 Quantum-Resistant Cryptography for Blockchains

Blockchains could have a two-phased approach for quantum readiness. First could be starting to adopt already-available quantum-resistant cryptography solutions. Second could be having a longer-term plan for not just being quantum-resistant but becoming truly post-quantum with a full implementation of post-quantum solutions. There are two main methods for full post-quantum cryptography, lattice-based cryptography and hash function-based cryptography, which are discussed in Chapter 8.

One strategy for interim quantum-resistant cryptography is implementing a widespread practice of incorporating non-reusable elements (signatures, keys, and addresses). This could be combined with methods that also confer full key protection. The current processes used in blockchains could be upgraded with more complex approaches that might be readily implementable and do not require further research. Most simply, a more complicated form of hash-based signatures could be used with an increasing number of bits that would be quantum-resistant, and would deliver full key protection. Other techniques involve more advanced elliptic curve cryptography methods such as super-singular elliptic curves (which extend over a field of endomorphic rings to convey a greater diversity of properties). The method could be implemented similarly to the way that elliptic curves are used today. Notably, Schnorr signatures (which are due to be implemented in an upcoming Bitcoin upgrade) are not quantum-resistant.

A medium-term strategy for quantum-resistant cryptography could be to instantiate computer science problems that are known to be hard for quantum computers to solve (unlike factoring and discrete logarithms) in cryptographic algorithms. Some examples include the Learning with Errors (LWE) problem (which requires finding a linear function) and multivariate polynomials (which require solving a system of multivariate quadratic polynomials over a finite field). Such solutions could provide a good roadmap for medium-term quantum-resistance.

Blockchain developers are aware of existing quantum-resistant solutions, but they have not been introduced because there is a trade-off between scalability and security, and quantum-resistant solutions are expensive in terms of system processing requirements. For example, Lamport signatures are known to be quantum-resistant, but are infeasible to implement in blockchains. Lamport signatures have a more complex structure than other signatures (the private key is produced by randomly generating $2*256$ numbers of 256 bits each, and the public key is hashes of these numbers). They create a secure one-time signature that is hard to break and are known to be quantum-resistant. However, Lamport signatures are not realistic for practical use in blockchains because storing them requires 200 times the space of the currently used ECDSA signatures (which are already stretching Bitcoin's scalability). With the ECDSA method, Bitcoin uses a 33-byte public key, and a maximum signature size of 73 bytes. Lamport signatures require much more space, 16 kibibytes of storage for public key data, and 8 kibibytes of storage for signatures (Kelly *et al.*, 2018, 13). Despite their quantum-resistivity, Lamport signatures do not make sense in the blockchain context. Hence, other quantum-resistant solutions are needed for the short term and robust post-quantum solutions for the long term.

References

Aaronson, S., Farhi, E., Gosset, D. *et al.* (2012). Quantum money. *Commun. ACM* 55(8):84–92.

Aggarwal, D., Brennen, G.K., Lee, T. *et al.* (2018). Quantum attacks on Bitcoin, and how to protect against them. *Ledger* 1(3):1–21.

Agnesi, C., Avesani, M., Stanco, A. *et al.* (2019). All-fiber self-compensating polarization encoder for quantum key distribution. *Opt. Lett.* 44(10): 2398–401.

Barker, E., Barker, W., Burr, W. *et al.* (2016). Recommendation for Key Management — Part 1: General (Revision 4). NIST Special Publication 800-57 Part 1 Revision 4.

Bennet, A.J. & Daryanoosh, S. (2019). Energy efficient mining on a quantum-enabled blockchain using light. arXiv:1902.09520 [quant-ph].

Bennett, C.H., Bessette, F., Brassard, G. *et al.* (1992). Experimental quantum cryptography. *J. Cryptol.* 5(1):3–28.

Blockchain.com (2019). Bitcoin Hash Rate. https://www.blockchain.com/en/charts/hash-rate. Accessed June 30, 2019.

Chen, L., Jordan, S., Liu, Y.-K. *et al.* (2016). Report on post-quantum cryptography. NIST Interagency Report 8105.

Dennis, E., Kitaev, A., Landahl, A. & Preskill, J. (2002). Topological quantum memory. *J. Math. Phys.* 43:4452–505.

Giles, M. (2018). The man turning China into a quantum superpower. *MIT Technol. Rev.*

Guan, J.Y., Arrazola, J.M., Amiri, R. *et al.* (2018). Experimental preparation and verification of quantum money. arXiv:1709.05882 [quant-ph].

Hasegawa, Y., Ikuta, R., Matsuda, N. *et al.* (2019). Experimental time-reversed adaptive Bell measurement towards all-photonic quantum repeaters. *Nat. Commun.* 10(378).

Hughes, R.J., Nordholt, J.E., McCabe, K.P. *et al.* (2010). Satellite-based quantum communications. In: *Proceedings of: Updating Quantum Cryptography and Communications 2010 (UQCC2010)*. Tokyo, Japan, October 18–20.

Ikeda, K. (2017). qBitcoin: A peer–to–peer quantum cash system. arXiv: 1708.04955 [q-fin.GN].

Jennewein, T., Simon, C., Weihs, G. *et al.* (2000). Quantum cryptography with entangled photons. *Phys. Rev. Lett.* 84(20):4729.

Kalinin, K.P. & Berloff, N.G. (2018). Blockchain platform with proof-of-work based on analog Hamiltonian optimisers. arXiv:1802.10091 [quant-ph].

Kelly, J., Lauer, M., Prinster, R. & Zhang, S. (2018). Investigation of blockchain network security: Exploration of consensus mechanisms and quantum vulnerabilities. MIT Course Syllabus.

Khan, I., Heim, B., Neuzner, A. & Marquardt, C. (2018). *Satellite-Based QKD.* OSA: Optics and Photonics News.

Kiktenko, E.O., Pozhar, N.O., Anufriev, M.N. *et al.* (2018). Quantum-secured blockchain. *Quantum Sci. Technol.* 3(3):035004:1–7.

Kiktenko, E.O., Pozhar, N.O., Duplinskiy, A.V. *et al.* (2017). Demonstration of a quantum key distribution network in urban fibre-optic communication lines. *Quantum Electron.* 47:798.

Kimble, H.J. (2008). The quantum internet. *Nature* 453:1023–30.

Lang, N. & Buchler, H.P. (2018). Strictly local one-dimensional topological quantum error correction with symmetry-constrained cellular automata. *SciPost Phys.* 4:007.

Le Gouet, J.L. & Moiseev, S. (2012). Quantum memory. *J Phys. B-At. Mol. Opt.* 45(12):120201.

Nakamoto, S. (2008). Bitcoin: A Peer-to-Peer Electronic Cash System. https://bitcoin.org/bitcoin.pdf. Accessed June 30, 2019.

Pant, M., Krovi, H., Towsley, D. *et al.* (2019). Routing entanglement in the quantum internet. *NPJ Quantum Inf.* 25:1–25.

Procopio, L.M., Moqanaki, A., Araujo, M. *et al.* (2015). Experimental superposition of orders of quantum gates. *Nat. Commun.* 6(7913):1–6.

Pugh, C.J., Kaiser, S., Bourgoin, J.-P. *et al.* (2017). Airborne demonstration of a quantum key distribution receiver payload. *Quantum Sci. Technol.* 2(2).

Rajan, D. & Visser, M. (2019). Quantum blockchain using entanglement in time. *Quantum Rep.* 1(2):1–9.

Sapaev, D., Bulychkov, F., Ablayev, A. *et al.* (2018). Quantum-assisted blockchain. arXiv:1802.06763 [quant-ph].

Sun, X., Wang, Q., Kulicki, P. & Zhao, X. (2019). Quantum-enhanced logic-based blockchain I: quantum honest-success byzantine agreement and qulogicoin. arXiv:1805.06768 [quant-ph].

Tessler, L. & Byrnes, T. (2018). Bitcoin and quantum computing. arXiv: 1711.04235 [quant-ph].

Wehner, S., Elkouss, D. & Hanson, R. (2018). Quantum internet: A vision for the road ahead. *Science* 362(6412):eaam9288.

Weihs, G., Jennewein, T., Simon, C. *et al.* (1998). Violation of Bell's inequality under strict Einstein locality conditions. *Phys. Rev. Lett.* 81(23):5039.

Wiesner, S. (1983). Conjugate coding. *SIGACT News*. 15(1):78–88.

Wilde, M.M. (2008). Quantum Coding with Entanglement. PhD Thesis. University of Southern California.

Chapter 7

Zero-Knowledge Proof Technology

Abstract

One of the biggest transformations underway in cryptography and blockchains is zero-knowledge proofs. The core idea is being able to prove validity without disclosing underlying information. Zero-knowledge proofs are a crucial enabling technology in the overall trend to make the internet more private and secure (zero-knowledge proofs are PrivacyTech and ProofTech). A zero-knowledge proof is a cryptographic method that separates data verification from the data itself. One party (the prover) can prove to another party (the verifier) the possession or existence of some information without revealing the information. Zero-knowledge proofs might be used in an extensive range of applications such as credit card authorization, account confirmation, and identity validation. Current state-of-the-art zero-knowledge proof systems include SNARKs, bulletproofs, STARKs, and Zether.

The movement of mathematical proof does not belong to the object, but rather is an activity external to the matter in hand

— G.W.F. Hegel (1807, §42, 24)

7.1 Zero-Knowledge Proofs: Basic Concept

Proof systems have been a research topic in cryptography and mathematics for a long time, and this work is now being seen in practical implementations in the blockchain ecosystem and beyond. Zero-knowledge proofs could be analogous to antivirus software for PCs, in the sense of being an early-stage problem-solving technology for a fledgling industry that becomes incorporated as a standard feature (while antivirus software combats malware, zero-knowledge proofs prevent hacking and snooping). The concept of a zero-knowledge proof was introduced as a proof that reveals no information except the correctness of the statement, in a landmark academic paper by Goldwasser *et al.* (1989). Zero-knowledge proofs are officially defined as "proofs that convey no additional knowledge other than the correctness of the proposition in question" (Goldwasser *et al.*, 1989). The researchers point out that the shortest answer to a proof is simply a one bit answer such as Yes/No or True/False. Most proofs are inefficient as a result of containing more information (knowledge) than the mere fact that the theorem is true. Hence, this chapter sets forth a computational complexity theory related to amount of knowledge that is required to be contained in a proof. In fact, having no knowledge (zero knowledge) of the underlying information is necessary in proofs, all that is necessary is having the 1-bit outcome indicating the truth value of the proof.

The paper highlights that the structure of many proofs is that they consist of a demonstration case (which contains excess knowledge), when the succinct output is simply a 1-bit Yes/No answer. An example is that to prove a graph is Hamiltonian, it suffices to exhibit a Hamiltonian tour in it, but this contains far more knowledge than the single bit indicating Hamiltonian or non-Hamiltonian. In graph theory, a Hamiltonian path is a traceable path in a directed or undirected graph that visits each vertex exactly once. A Hamiltonian cycle (or Hamiltonian circuit) is a Hamiltonian path that is a cycle. The method of using such cycles or circuits that indicate a path through a landscape (i.e. a demonstration case) is reflected in computational cryptographic circuit design. Although the path provides a demonstration case of the proof, it is inefficient because the excess knowledge is unnecessary when the salient output is a 1-bit answer.

The Hamiltonian path method is a proof strategy; the logic is that if a demonstration case exists, the theorem must be true. However, in another proof strategy, the output could be more efficiently reduced to a 1-bit answer, and this is the key take-away for computational proof design and online security. The idea is to construct zero-knowledge proofs (proofs with zero excess knowledge) that have the most efficient output possible, such as a 1-bit answer. The further point that the underlying theorem or information itself is kept private in the proof output, because it does not matter to the efficiency of the proof result, is a side benefit which has become part of the essential feature set of zero-knowledge proofs in blockchains being both a private and efficient computational proof technology.

7.2 Zero-Knowledge Proofs and Public Key Infrastructure Cryptography

7.2.1 *Public key infrastructure*

The significance of zero-knowledge proofs is that they could potentially become a standard part of cryptographic network infrastructure. The problem is that the internet was not designed to be secure, and hacking, break-ins, and espionage have become routine. Zero-knowledge proofs are one way of making the internet secure and private by ensuring the confidentiality and integrity of data. The internet can be turned into a VPN (virtual private network) such that all transactions are private. Zero-knowledge proofs are a privacy overlay for the internet, and are based on public key infrastructure (PKI).

PKI is a method for securely exchanging cryptographic keys over a public channel. Contemporary PKI infrastructure was originally conceptualized by Merkle (1978) and implemented as the Diffie–Hellman key exchange protocol (Diffie & Hellman, 1976). The open internet is a public channel. Users would like to send each other messages across the public channel, and need a way to encrypt them so they stay secure and cannot be read by other parties. (In computer science, this is called the Byzantine general's problem, namely that in a war, generals would like to send secure messages to each other across the battlefield.) In a PKI system, each user has a public–private key pair. Users keep their private key secure and knowable only to themselves, and distribute their public key on the open internet. The keys are each strings of a certain length (for example,

32-character alphanumeric codes) that are mathematically related (for example, comprising two points on a certain elliptical function curve).

With a clever scheme using public keys, users can send each other encrypted messages over the internet such that only the receiving party can decode and read them. If A and B are both using the same PKI infrastructure, if A wants to send B a secret message, A asks B for B's public key. B's key might be published on the internet, or B sends A the public key across the open internet. A uses B's public key to encode a message to B (with the PKI software) that only B can read, and sends the encoded message. B receives the encoded message, and because only B has B's private key, B is able to decode the message and read it. This is possible because B's public and private keys are related in the encryption mechanism; any message encoded with B's public key can be decoded with B's private key. A and B are participating in the same PKI system, meaning that each is running the same software which mathematically relates the public and private key pairs held by each specific user together in a known method to encrypt and decrypt the messages on either end without revealing the underlying private keys.

7.2.2 *Blockchain addresses*

Blockchain addresses (used by Bitcoin and many other platforms) are a slightly more complicated version of PKI cryptography, in that both the public and private keys are generally hidden from open distribution on the internet. The address is a hash of the private key and the public key with some additional information and encoding, and is shared openly on the internet. A unique address is typically created for each transaction. In blockchain distributed ledgers, it is not B who is receiving the message, it is the blockchain peer-to-peer network. The message that is sent is broadcast as an encrypted transaction from A's wallet. The network, meaning all of the machines that are running the software, each receive, validate, and process the transaction. Specifically, the blockchain network node software automatically decrypts the message and validates the transaction (checking A's address and if the transaction amount matches the available coins, then checking B's address). If the transaction is valid, it is passed into a pool of unconfirmed transactions to be included in a block. The point is that the peer-to-peer network acts in the role of receiving and decrypting the message from A, and A's private key remains undisclosed.

7.3 Zero-Knowledge Proofs: Interactive Proofs

Zero-knowledge proofs are a further extension of the basic concepts used in PKI cryptography. The proofs rely on the fact that each user has a private key that is only accessible to that entity (whether a person, computer, robot, or IoT sensor). Since each entity has their own private key, encrypted messages can be sent to them that only they can decrypt and answer. It is the same idea as sending messages that only the recipient with the private key can decode. This functionality can be used to test identity claims that parties are really who they say they are. Demonstrating the truth claim requires an iterative interactive process.

To prove A's identity, B selects some text to send to A as a secret message (such as "secret message"). B encrypts the message with A's public key and sends it to A. If A is really A, then A has A's private key and can decrypt the message and pass it back to B as an open string on the internet. A sends back "secret message". However, if A is an imposter, A might be able to guess the message contents, so the process has to repeat a sufficient number of times for B to be convinced that A is not randomly guessing the secret message text. There is an iterated interaction (hence the name interactive proofs) such that A can answer a sufficiently-high probabilistic amount of times (10–20) attesting that A is really A. Then, when B is sure that A is really A, B sends the real transaction. The key point is that A does not share any private information (the private key), but can prove its identity to B. Notably, A and B do not need to know or trust each other.

There are important implications of using zero-knowledge proofs for internet data privacy. In a zero-knowledge proof-enabled world, A's identity can be authenticated and verified without A having to share actual information (whether it is a private key, social security number, account number, or other directly-identifying information). Blockchains assume lack of trust from the beginning, and build a computational system to confer trust. Zero-knowledge proofs demonstrate this principle.

7.3.1 *Interactive proofs: Graph isomorphism example*

Zero-knowledge proofs are a known technique outside of the blockchain context. A zero-knowledge proof is a proof in which the verifier of the proof does not learn anything about the statement in question except that it is true. A canonical example of zero-knowledge proofs is the graph isomorphism problem. Two graphs are isomorphic if the vertices can be relabeled in one of the graphs to obtain the other. (For example, if vertex A is connected to vertex B in one of the graphs, then vertex A is likewise connected to vertex B in the other graph.) More formally, the graph isomorphism problem is the situation of determining whether it is possible to permute the vertices of one graph so that it is identical to another graph.

Considering two graphs, each with 10 vertices, it is not clear just by looking at them if they are isomorphic. However, if the vertices can be labeled 1–10 on the first graph, and 1–10 on the second graph, it can be checked computationally whether the two graphs are isomorphic. Notably, whereas solving graph isomorphism problems with classical computers is difficult, quantum computers are thought to be able to easily solve such problems. The method is creating the graph G, and calculating the superposition over all vertex labelings of G to test for isomorphism. One upshot is that a certain method of cryptography, graph isomorphism-based cryptography, is not quantum-resistant.

The graph isomorphism problem introduces the concept of one-way functions. On the one hand, it is easy to check that the two graphs are isomorphic (if this claim is being made), but on the other hand, it is hard to derive the actual isomorphism (the labeling). In a one-way function, similarly, it is easy to verify the claim that is being made, but difficult to calculate the inputs that constitute the claim.

The issue that interactive proofs try to solve is as follows. Considering two graphs, if the graphs are isomorphic, that is easy to prove. The prover makes a claim that two graphs are isomorphic, and gives the labeling to the verifier. The verifier can easily check whether this is an accurate isomorphic mapping between the two graphs, and if it is, then indeed, the graphs are isomorphic. However, if the graphs are non-isomorphic, this is hard to prove. The problem is how the prover can demonstrate the claim to the verifier that the graphs are non-isomorphic.

Such claims can be demonstrated with an interactive zero-knowledge proof. The verifier picks one of the two graphs at random, creates a small change or permutation to the graph (such as a slight reordering of the data points), and asks the prover to identify which graph it is (A or B). If the claim is correct, the prover should be able to consistently answer the question correctly. If the claim is incorrect, the prover should only be able to answer the question correctly 50% of the time (a 50/50 chance of being correct). Therefore, after a few rounds, it should be easy for the verifier to confirm whether or not the prover is making a truthful claim.

The key point about the zero-knowledge proof is that the verifier becomes convinced that the graphs are non-isomorphic without learning anything specific about the underlying data in the graphs (i.e. keeping the underlying data of the vertex labeling private). This is the core idea in zero-knowledge proof technology that the prover can prove that they have access to their own private key, without sharing that key, by generating messages with the key that the verifier can confirm.

In terms of computational complexity, assuming (non-trivially) that one-way functions exist, it can be shown that zero-knowledge proofs exist for every NP-complete problem (Goldreich *et al.*, 1991). The implication is that as a computational complexity class, all NP-complete problems are reducible to each other, in the sense that it is possible to give a zero-knowledge proof for any one NP-complete problem. This suggests the treatment of computational complexity by class.

7.3.1.1 *ZKP example: Colorblind and red and green balls*

In this example of zero-knowledge proofs, the problem is how a person who is not colorblind can prove this fact to a person who is colorblind. The zero-knowledge proof solution is as follows. The colorblind person holds two balls, red and green. The colorblind person puts their hands behind their back, decides whether to switch the balls or not, and then reveals the balls in their hands. The non-colorblind person can immediately tell if the balls have been switched and say "switched" or "not switched" (in the form of a 1-bit answer). To make sure this was not a lucky guess, the rounds continue until there is probabilistic certainty that the non-colorblind person's claim is true, that indeed she is not colorblind, and not just randomly guessing whether the balls have changed hands. The fact of non-colorblindness can be proved without sharing underlying data about the functioning of the visual receptors, the "private information" in this case.

There are two levels of information in the system, the underlying information about the color of the balls, red and green, and the meta information of whether there has been a switch in any round. The non-colorblind person can distinguish and report publicly whether there has been a switch without revealing the underlying information itself (red and green). Zero-knowledge proofs operate on this principle of two-tier information systems comprising the underlying information itself and statements that can be made about the information without disclosing the information.

7.3.1.2 *ZKP example: Grocery store private transactions*

A basic use case of zero-knowledge proofs in blockchains is the grocery store example. Zero-knowledge proof technology could be used to enable private purchases in the sense of not having to disclose an actual credit card number when making a purchase. Someone could go to the supermarket and purchase goods with a credit card without revealing the card information to the merchant. A smart credit card using zero-knowledge proof technology could identify the cardholder to the merchant without giving the merchant the knowledge (the credit card number). The cardholder's private information is protected, and the merchant is also protected, in the zero-knowledge proof that this is a valid credit card with the available limits, and not a forged card.

7.4 Zero-Knowledge Proofs in Blockchains

The main method for enabling private transactions in blockchains is zero-knowledge proofs. The challenge with blockchains is the publicly-verifiable nature of the technology. The conundrum is how transactions can be posted to the public ledger in a shared trustable (because it is trustless, meaning computationally-derived trust) method that nevertheless does not disclose the specific details of the transaction. In the basic instantiation of blockchains (i.e. Bitcoin), the public ledger tracks sender address, recipient address, and transaction amount. With addresses and transaction amounts being disclosed, blockchain analytics companies and other parties have been able to assemble transactions and link overall balances of asset ownership to real-life personal identities. In recent years, there has been a move to implement private transactions in both public and enterprise blockchains in which sender and recipient address, and transaction amount, are masked or shielded in the public ledger, and in the consensus process. Privacy-protected transactions might be confidential (shielding the amount that is being transferred), anonymous transactions (shielding the addresses that indicate who is transferring to whom), or both. Although blockchain addresses provide some privacy due to their pseudonymous nature (addresses are typically 32-character alphanumeric codes), if exchanged in the open, they are possibly traceable in certain ways.

7.4.1 *Zero-knowledge proofs: Range proofs*

The central problem is making transactions completely private, while still allowing them to be externally-validated by the blockchain system. Due to the structure of blockchain transactions, it is possible to use a special kind of zero-knowledge proof called a range proof to prove that the transaction amount is positive and within a certain narrow range, without disclosing the amount. Range proofs indicate that the amounts in the inputs and the outputs of the transaction are equal, and add up to zero, without having to disclose what the amounts are. In confidential transactions in blockchains that shield the transaction amount, range proofs are the kind of zero-knowledge proofs that are typically used.

7.4.2 *Unspent transaction outputs model*

The canonical transaction model of blockchain systems, used by Bitcoin and other blockchains, is the unspent transaction outputs (UTXO) model. At any snapshot in time, a user's wallet indicates the UTXOs that the user owns. These are the outputs of previous transactions that the owner received as money that was transferred into the user's wallet from other parties. The UTXOs are an inventory of the digital money that is available to be spent. At any moment, the Bitcoin ledger consists of all outstanding UTXOs that are open cash balances that can be spent, and the history of all transactions to date that previously used these coins. Unlike a dollar bill, a Bitcoin carries its full history since inception with it. The UTXO system means that every Bitcoin can be traced back over its history since the start of the monetary system.

When the end user goes to execute a transaction, the wallet locates UTXOs (i.e. available cash) totaling the amount needed for the transaction (any residual amount not needed for the transaction is returned to the wallet as another UTXO for future transactions, like receiving change from a cash transaction). The way that the UTXO system is structured allows various checks and balances to be performed, which is how the blockchain operates. The blockchain software automatically performs a real-time check to verify that the UTXOs submitted in a transaction are indeed owned by this particular wallet's public-private key pair, and are available to be spent (they have not been spent in a recent transaction by this owner). The wallet software performs a validity check of the transaction such that the sum of the transaction outputs does not exceed the sum of the transaction inputs (it is not possible to spend more than you have). The UTXO input–output checks in the transaction are essentially a more complicated version of the checksum concept. Checksums are widely used to ensure data integrity during file transfer or access from storage (most basically that the sum of the bits received is equal to sum of the bits sent).

In reality, the checks made by the blockchain software are a bit more complicated. To validate a Bitcoin transaction, a miner or full node checks that the signatures are correct (validating ownership), the inputs are unspent, the sum of the inputs is equal to the sum of the outputs plus the fees, and that all of the outputs are positive. However, in conventional blockchain transactions, the transaction amount is visible as public information that is tracked by analytics companies. As a privacy improvement, confidential transactions are proposed to shield the transaction amounts in Bitcoin transactions (Maxwell, 2016). In a confidential transaction, the value is encrypted so that it is not possible to see how much is being transacted. The question is, how can the mining operation validate the transaction and prevent the money from being double-spent if the amount is encrypted?

The way that transactions can be private, while still allowing for external validation is through range proofs. Range proofs are based on sigma protocols, which are protocols (instructions) for small-scale non-interactive zero-knowledge proofs. The key idea is that it is possible to encrypt the values in the inputs, and in the outputs, and apply a proof that indicates that they cancel each other out. The sum is zero, so it is confirmed that there is an equal amount of inputs and outputs, and that no new money is being created (money is not being double-spent).

The first step in executing the range proof is replacing the transaction amount with a commitment (a cryptographic building block called a Pedersen commitment is used). Privately, the transaction issuer remains committed to the exact amount, but publicly it is only seen that a commitment exists, not the amount of the commitment.

The second step in the range proof is performing a non-interactive zero-knowledge proof. The prover (the account owner) uses a common reference string (analogous to a public key) that is available to both the prover and the verifier, and creates a proof that the committed value is positive and within a certain range necessary for the transaction. (The account owner's wallet does this automatically.) The verifier (the counterparty to the transaction, but really a function performed by the blockchain software, which is publicly-verifiable by any external party) then checks the proof for this commitment, and becomes convinced that the committed value is positive and within the necessary range for the transaction. It is a zero-knowledge proof, in that the verifier does not learn any information other than the fact that the committed value is positive and within a certain range. Most importantly, the verifier does not learn the precise amount. The prover proves that the committed value is within some range, and then a proof of that statement is executed.

The proof structure of Succinct Non-interactive Arguments of Knowledge (SNARKs), the first form of zero-knowledge proofs used in blockchains, uses the following mechanisms. The prover makes a claim to having UTXOs available in their wallet to use in the current transaction by proving that they know a Merkle tree path to this particular unspent coin in the blockchain ledger. The transaction proof consists of confirming that the Merkle tree paths of the inputs are equal to the Merkle tree paths of the outputs. This is how Bitcoin processing already works, the new element is making a proof of this, while keeping the amounts hidden.

7.5 State-of-the-Art: SNARKs, Bulletproofs, and STARKs

There are three main kinds of zero-knowledge proof technologies being deployed in blockchains: SNARKs (2014), bulletproofs (2018), and STARKs (2018) (Ben-Sasson *et al.*, 2014, 2018; Bunz *et al.*, 2018). Each has certain advantages and disadvantages (Table 7.1). The three key trade-off parameters are the size of the proof (which is of concern for quick transfer and because it is stored in the blockchain), and the time it takes to generate the proof and to verify the proof. The proof time and verification time values in the figure (fast, very-fast, and not very-fast) are vague distinctions that should not be taken literally, but rather as a general indication of the relative strengths of the different proof systems. Whereas the initial technology, SNARKs, requires a trusted setup, more recent proof systems alleviate this need, and are called transparent, meaning that the process itself is transparent and does not require a trusted setup to install the proof circuit in the first place before others can use it on an ongoing basis.

Table 7.1. Comparison of zero-knowledge proof systems.

ZKP system	Proof size	Trusted setup required?	Proof time	Verification time	Post-quantum secure?
SNARKs	1.3 kB (sapling)	Yes	Fast	Fast	No
Bulletproofs	1–2 kB	No	Fast	Not very-fast	No
STARKs	20–30 kB (was 200 kB)	No	Not very-fast	Very-fast	Yes

7.5.1 *SNARKs and multi-party computation*

The original blockchain zero-knowledge proof system is SNARKs. SNARKs have the shortest proof length, and verification time is short. However, a trusted setup is required using multi-party computation in a public parameter generation ceremony to establish proof circuits. Every time the project wants to make a change to the zero-knowledge proof circuit or instantiate a new circuit, there needs to be a trusted setup process. SNARK setup uses multi-party computation, which is an advanced method for secure blockchain computing derived from multi-party cryptography protocols. In this approach, multiple non-trusting computers conduct a computation on their own unique fragments of a larger dataset to collectively produce a common outcome, without any one node knowing the details of the fragments held by the others.

One of the main applications of multi-party computation is secure key generation and management in the context of facilitating group operations. The technique is generically useful and features prominently in the trusted setup process used in SNARKs. The idea is that it is possible to share signing responsibility among a group of otherwise non-trusting entities. Key generation is part of the trusted setup, in that the private key that executes the transaction is a collectively-generated value by the participants, which is the security feature that makes the process invulnerable to attack.

7.5.2 *Bulletproofs and STARKs*

After SNARKs, the next notable blockchain proof technology is bulletproofs, which are short non-interactive zero-knowledge proofs that do not require a trusted setup (bulletproofs: proofs that are very small and fast (like a bullet)). The main advantage of bulletproofs over SNARKs is that there is no trusted setup.

The third important zero-knowledge proof technology in blockchains is STARKs (Scalable Transparent [no trusted setup] Arguments of Knowledge). STARKs are an improved version of SNARKs without a trusted setup, and have a longer proof time, but a very fast verification time. In terms of size, SNARKs are very small, bulletproofs are ~1–2 kilobytes, and STARKs are currently 20–30 kilobytes (down from over 200 kilobytes). Bulletproofs have a larger proof size than SNARKs by a factor of ten, being about 1–2 kilobytes. STARKs are another order of magnitude larger than bulletproofs (about 20 to 30 kilobytes).

The main advantage of STARKs over bulletproofs is scalability. In STARKs, the verification time is much smaller than the time needed to run the computation. Bulletproofs are a valid privacy solution, but do not offer scalability benefits because the time needed to verify a computation is at least the size of the computation, whereas in STARKs, it is exponentially smaller.

To give an indication of the trade-offs of different proof technologies, STARKs might also be compared with ZKBoo (faster zero knowledge for Boolean circuits) (Giacomelli *et al.*, 2016). ZKBoo has a different set of parameters such that the proof time is fast, but the proof size is large. The point is that the different proof systems are more conducive to certain applications. For fast end user execution with high through-put volume, and when the standard circuit does not need to change, SNARKs are the most efficient. Bulletproofs are more flexible in setup, but not as efficient as SNARKs. STARKs are for large provers acting as a service infrastructure for the rest of the blockchain system, who need to prove claims on a regular basis such as solvency status, where the proof time can be longer, but the verification time is very fast.

STARKs are intended more for the "enterprise proof market," whereas SNARKs are directed at the "consumer proof market". STARKs are aimed at institutional customers who need zero-knowledge proofs to prove claims about themselves to the market. This includes solvency proofs and other kinds of proofs regarding the general accountability and auditability of institutions, including banks, exchanges, other financial institutions, and also stablecoins. Solvency proofs and auditability proofs could become a standard feature of the blockchain ecosystem, similar in function to a credit rating or a Dun & Bradstreet check for new vendors. STARKs are also aimed at internal operations at institutional clients, for example at decentralized exchanges, for use in settlement, order book management, and fraud prevention (such as preventing front-running).

Regarding whether zero-knowledge proof systems are post-quantum secure, it depends on the cryptography used in the security on which a particular zero-knowledge proof system rests. The earlier projects such as SNARKs and bulletproofs are not post-quantum secure because they rely on cryptographic assumptions which are known to be vulnerable to quantum attacks (namely the hardness of discrete log calculations in certain elliptical curve groups). On the other hand, zero-knowledge proof systems that only use collision-resistant hash functions are plausibly post-quantum secure. This includes STARKs and also Hyrax and Aurora.

There are many different zero-knowledge proof projects. The field is growing and there are many ongoing developments. One example is Bolt, which implements zero-knowledge proofs in payment channels on the Lightning Network (a Layer 2 payment channel overlay to Bitcoin) (Green & Miers, 2017). Substantial change may be likely as the technology matures. There are various known computational methods that prove zero-knowledge in various domains, such as Schnorr protocols, sigma protocols, and password hashes. However, private transactions on blockchains require more complicated and dedicated forms of zero-knowledge proof technology such as SNARKs, bulletproofs, STARKs, and Zether.

7.6 State-of-the-Art: Zether for Account-Based Blockchains

Zether is a zero-knowledge proof solution for account-based models and smart contract platforms, as opposed to other zero-knowledge proof systems that are for use in UTXO models (Bunz *et al.*, 2019). Zether is a variation of bulletproofs (Bunz *et al.*, 2018) that adds additional functionality to extend confidential transactions to account-based models. A comparison of transaction systems is illustrated in Table 7.2.

Zether combines bulletproofs with sigma protocols (into "sigma bullets"). Sigma protocols are a standardized construction for zero-knowledge proofs with a three-step structure of commitment, challenge, and response. On their own, sigma protocols are prohibitively large for implementing a system of confidential transactions in blockchains and require additional

Table 7.2. Transaction systems comparison: Confidentiality and anonymity.

Transaction system	Cryptocurrency project	Confidential (amount shielded)	Anonymous (address shielded)
UTXO model	Bitcoin	—	—
UTXO model	Monero	X	—
UTXO model	MimbleWimble: Grin, BEAM	X	—
UTXO model	Zcash	X	X
Account model	Ethereum		
Account model	Zether	X	X

manipulation. The name Zether connotes a privacy-protected version of Ether (zero-knowledge Ether), the native cryptocurrency of Ethereum, the largest account-based cryptocurrency and smart contract platform. Although designed for Ethereum, and implemented as an Ethereum smart contract, Zether is generally applicable to any blockchain with an account-based system.

The transaction model used by cryptocurrency projects is either the UTXO-based model or the account-based model. Bitcoin as well as privacy cryptocurrencies Monero and Zcash use the UTXO-based model. Ethereum uses the account-based model. There are trade-offs to each model, in general, and especially when implementing zero-knowledge proof technology. On the one hand, a benefit of account-based models is that they are more scalable because the state of the account can be compressed. One account that receives a million transactions in Bitcoin (such as an exchange account) would take up a lot of space to represent, whereas in Ethereum, would only take a few bits. On the other hand, it is more difficult to process account-based models because there are more variables in computing the state (they are more like a bank account). As a result, new functionality such as zero-knowledge proofs is harder to implement in the account-based model.

7.6.1 *Bulletproofs: Confidential transactions for UTXO chains*

The basic structure of a bulletproof is that a party opens a commitment, and then proves that the commitment has some value (a positive value within a small range).

In the general use case of bulletproofs for creating confidential transactions, two cryptographic building blocks are employed, commitments and zero-knowledge proofs. Instead of the transaction amount being broadcast publicly as in Bitcoin and Ethereum transactions, the balance is replaced by a commitment. Bulletproofs use Pedersen commitments. The commitment has two properties: it shields the amount, and it is binding (once there is a commitment to a value, it cannot be opened to another value later). It can be seen publicly that a commitment has been made, but only the sending party knows the amount, which has been committed. The receiver does not know (for sure) the transaction amount until the commitment is opened.

The challenge with confidential transaction is how the blockchain system can check to validate the transaction and prevent double-spending given that the amount is hidden. Zero-knowledge proofs can be used to check various conditions such as the transaction inputs are equal to the outputs plus the fees. In particular, the range proof functionality of zero-knowledge proofs is engaged to prove that a commitment is in a small range.

7.6.2 *Zether: Confidential transactions for account chains*

Zether combines bulletproofs with some of the additional cryptographic complexity provided by sigma protocols to operate in an account-based transaction model. Transaction execution is much easier in a UTXO-based model than in an account-based model. The state of a UTXO balance is binary, a certain UTXO is either spent or unspent. A computational system such as a blockchain can quickly check the state of a UTXO balance to determine if it is available so that it is not being double-spent. The state of an account is much more complicated. Consider a bank account in which various kinds of payments may be coming into and going out of the account.

Bulletproofs cannot be directly applied to the account-based transaction model. An issue arises in the commitment structure of bulletproofs of replacing the transaction amount with a commitment. Since the account state is used in the computation to generate the commitment, the commitment will be invalid if the account state has changed between the time of the initial commitment and the time the commitment is recomputed for the transaction execution. Account states may change frequently, whereas a UTXO's state remains "unspent" until the coin is "spent" by a particular address.

To accommodate the situation of changing account states, sigma bullets (bulletproofs + sigma protocols) implement a paired account system of a stable account and a temporary account. The stable account is the account from which commitments and proofs are made, and the temporary account is used for the incoming and outgoing activity of the account. There are periodic rollovers (every few blocks) to merge the temporary account into the stable account. To prove that there is no double-spending between the linked accounts, a basic zero-knowledge proof is performed by hashing the private key to create a proof that the hash of the private key is correct (Monero and Zcash also use this basic application of zero-knowledge proof technology to confirm the private key). The stable–temporary account pairs provide a protected time window for the validity of commitments to be updated.

The commitment changes (and needs to be re-proven as valid) if the account state changes, and also if the account holder adjusts the commitment. Although Pedersen commitments are binding, there may be some flexibility for the account holder to add to the commitment or make other kinds of allowable adjustments. If the account holder changes the commitment, another zero-knowledge proof must be submitted that proves that the requisite commitment for the initial transaction is still in place. Formally, Zether combines bulletproofs-based range proofs with sigma protocol-enabling features (specifically ElGamal encryptions) to provide zero-knowledge proofs in the case of updates to commitments.

Beyond being able to execute zero-knowledge proofs in account-based transaction models, there are other potential scalability benefits to these proofs. Zether's paired account model with periodic rollovers between the temporary and the stable accounts offers greater scalability than the UTXO model. Whenever the account rolls over, the set of proof nonces can be deleted, whereas they are stored in the ever-growing state in UTXO models such as Monero and Zcash. The UTXO set continues to grow as the coin is spent from party to party. Not only are the proof nonces included in the growing UTXO set, but it is also necessary to store nullifiers to prevent double-spending. In the account-based model of proofs, however, there is no state that continues to grow as transactions are added, and the overall proof structure may be more scalable for the longer term.

7.6.3 *Confidential smart contract transactions*

Bulletproofs are confidential transactions for the UTXO-based model. Zether, in the basic case, is confidential transactions for the account-based model. However, account-based models are used for more than just one-off transactions between parties, they are also used for smart contracts (more complicated transactions taking place over time). The next step is extending the functionality so that smart contracts themselves can use confidential amounts. The issue is that smart contracts only have public states (there is no private state that is the whole point of blockchain transparency). Hence, if a smart contract is to use a confidential amount, it must be paired with another smart contract that enacts the confidentiality. Similar to the paired account structure between the stable and temporary accounts for managing account state changes in Zether, a paired contract is set up with an auxiliary contract to carry out the confidentiality shielding of the amount in the usual way. To prevent double-spending or other malfeasance, the contract amount is locked in the other contract in an escrow-type fashion, and unlocked when the transaction is complete.

Confidential smart contract transactions enable a new set of applications. These include confidential staked-voting (in which a party's vote is proportional to the amount of stake, but kept confidential). New kinds of auction functionality are possible such as auctions without collateral and auctions with perfect bid privacy (previously the pledged collateral would effectively reveal an upper bound on the bid). With a Zether-based auction, bids could be submitted and then when the auction concludes, parties open their bids and the winner proves that their bid is higher than the last highest bid. Another application supporting Layer 2 developments is confidential payment channels (which are all smart contract-based). Further, confidential proof-of-stake systems are enabled that do not reveal the public key of the voting party to stake the vote as is the current situation. This would contribute substantially to making existing blockchain structures quantum-secure.

7.6.4 *IPFS interactive proof-of-time and proof-of-space*

The proofs used in the IPFS/Protocol Labs system are in the same structure as SNARKs. They involve a computational method for executing interactive proofs. The first use case is proving safe decentralized storage. The necessary proofs are proofs of having performed good player behavior (providing a safe storage resource) in the recent past. Decentralized storage providers need to prove consistently that they are storing useful bits (i.e. real files, not just random bits) over time (so that the files can be safely accessed in the future). Storage providers need to provide a proof-of-space and a proof-of-time (using space to store real files over time). For the system to be trustworthy and secure, the storage providers have to prove almost constantly, for example, at a rate of one hundred proofs throughout the day.

This quickly becomes inefficient, so the innovation is to have the provers do their own proving (cryptographically), and send a daily Merkle root that corresponds to all of the proofs (a proof of proofs). Time-stamping indicates the proofs have actually occurred on a regular basis. The verifier (miner) then confirms the proof of proofs from the Merkle root. The proofs are aggregated and compressed into the Merkle root, but they are all linked and callable through the hash tree structure, just as all 580,000 blocks of 10 years-worth of Bitcoin blockchain transactions can be compressed and linked into one Merkle root. The implication is that verification could become a standard mechanism in cryptographic technologies. The IPFS solution, called a proof-of-replication (PoRep), is an interactive proof system in which a prover defends a publicly verifiable claim that it is dedicating unique resources to storing one or more retrievable replicas of a data file (Fisch, 2018).

The IPFS proof system is notable as a complex yet toy model of what is essentially the same structure of proofs in STARKs (Chapter 8). The general strategy is to construct a proof structure that has a vast and elaborate apparatus of hash functions, Merkle paths to hashes, and time-stamped proofs that can be quickly checked for inconsistency to verify their validity. The proofs are interactive (computationally interactive) in that the computational processes take place over time. This introduces a dimension of time complexity into the proof structure, which makes it even more difficult for malicious agents to submit an inaccurate proof.

An additional innovation is using slow-time hash functions, which are designed to be slow and inefficient to deter malicious players (bona fide storage providers provide persistent file storage and are not bothered by hashing functions that operate in slow-time). The idea is similar to Bitcoin blocks specifically taking about 10 min to confirm in order to enable enough peer-based miners worldwide to examine and confirm the transactions. Typically, the goal would be having quick and efficient fast-hashes for cryptographic operations such as authentication (message authentication, digital signatures, transaction confirmation) and data integrity checks during file transfers (checksums). However, slow-hashes are useful for other operations, including the game theoretic principles of deterring malicious player behavior. Fast and slow hashing is another example of time complexity, and using computational complexity as a technology design principle.

References

Ben-Sasson, E., Bentov, I., Horesh, Y. & Riabzev, M. (2018). Scalable, transparent, and post-quantum secure computational integrity. ia.cr/2018/046.

Ben-Sasson, E., Chiesa, A., Garman, C. *et al.* (2014). Zerocash: Decentralized anonymous payments from Bitcoin. In: *Proceedings of the IEEE Symposium on Security & Privacy*. Oakland, pp. 459–74.

Bunz, B., Agrawal, S., Zamani, M. & Boneh, D. (2019). Zether: Towards privacy in a smart contract world. ia.cr/2019/191.

Bunz, B., Bootle, J., Boneh, D. *et al.* (2018). Bulletproofs: Short proofs for confidential transactions and more. In: *39th IEEE Symposium on Security and Privacy 2018.*

Diffie, W. & Hellman, M.E. (1976). New directions in cryptography. *IEEE Trans. Inf. Theory.* 22(6):644–54.

Fisch, B. (2018). PoReps: Proofs of space on useful data. ia.cr/2018/678.

Giacomelli, I., Madsen, J. & Orlandi, C. (2016). ZKBoo: Faster zero-knowledge for Boolean circuits. In: *25th USENIX Security Symposium.* Austin, TX, August 10–12, pp. 1069–83.

Goldreich, O., Micali, S. & Wigderson, A. (1991). Proofs that yield nothing but their validity or all languages in NP have zero-knowledge proof systems. *JACM* 38(1):691–729.

Goldwasser, S., Micali, S. & Rackoff, C. (1989). The knowledge complexity of interactive proof systems. *SIAM J. Comput.* 18(1):186–208.

Green, M. & Miers, I. (2017). Bolt: Anonymous payment channels for decentralized currencies. In: *CCS '17 Proceedings of the 2017 ACM SIGSAC Conference on Computer and Communications Security.* Dallas, TX, October 30–November 03, 2017, pp. 473–89.

Hegel, G.W.F. (1807). *Phenomenology of Spirit.* Trans. A.V. Miller. Oxford, UK: Oxford University Press.

Maxwell, G. (2016). Confidential transactions. https://people.xiph.org/~greg/confidential_values.txt. Accessed June 30, 2019.

Merkle, R.C. (1978). Secure communications over insecure channels. *Commun. ACM* 21(4):294–9.

Chapter 8

Post-quantum Cryptography and Quantum Proofs

Abstract

Zero-knowledge proofs are just the basic concept in proof technology. Many other sophisticated formulations of proofs and related quantum-secure technology are proposed. Earlier forms of zero-knowledge proof technology are not quantum-resistant. Proofs based on generic hash functions as opposed to classical PKI cryptography are thought to be post-quantum secure because the cryptography can be upgraded, whereas classical PKI cryptography is known not to be quantum-secure. STARKs are a new form of zero-knowledge proofs for blockchains with a sophisticated architecture based on error correction codes, random queries, and inconsistency checks. The two main approaches to post-quantum cryptography are lattice-based cryptography and hash function-based cryptography. Lattice-based cryptography is currently the main method being promulgated in next-generation US NIST algorithm development. Keeping public digital infrastructure safe is a key concern. Post-quantum cryptography suggests various approaches to prepare cryptographic infrastructure for a potential future of quantum computing.

8.1 STARKs

As introduced in Chapter 7, the three main kinds of zero-knowledge proof technologies being deployed in blockchains are SNARKs (2014),

Bulletproofs (2018), and STARKs (2018), developed by two research teams (Ben-Sasson *et al.*, 2014, 2018; Bunz *et al.*, 2018).

STARKs refer to a transparent zero-knowledge proof system with very-fast proof verification time (in which verification scales exponentially faster than database size). The "T" stands for transparent, indicating that a trusted setup is not required, thereby resolving one of the primary weaknesses of SNARKs. STARKs are intended for scalability, such that the verifier can discern between true and false statements exponentially faster than it would take to run the computation, which is a substantial improvement compared to other methods. STARKs also have simpler cryptographic assumptions, which avoid the need for classical formulations such as elliptic curves, key pairings, and the knowledge-of-exponent assumption, and instead use hash functions and information theory. The implication of proof systems that rely only on hash functions and information theory (and not on classical PKI cryptography) is that they would be post-quantum secure.

The trade-off of very-fast verification and no trusted setup is a larger proof size. The STARKs formulation at present has a one order of magnitude larger proof size than bulletproofs for a single shielded transaction (about 20–30 kB, as compared with 1–2 kB for bulletproofs). For classes of applications that require a high degree of trustless trust, the cost of the larger proof size may be worth it. The STARK solution is designed for public blockchain infrastructure providers and applications such as decentralized exchanges that need to demonstrate solvency proofs and other zero-knowledge provable conditions on a regular basis. For these applications, a fast verification time is important, but not necessarily a fast proof generation time.

8.1.1 *Proof technology: The math behind STARKs*

STARKs use cryptographic hash functions (modeled as random oracles), and represent an advance in the trajectory of probabilistically checkable proofs (PCPs) and interactive oracle proofs (IOPs). The main source of the advance, improved performance in the form of very-fast verification time, is called scalable transparent IOPs of knowledge. To realize the advance, a known method, the fast Reed–Solomon IOP of proximity protocol (Ben-Sasson *et al.*, 2017), is used to generate a procedure for an exponential speedup in verification time, using an error-correction code-based method (Ben-Sasson *et al.*, 2018, p. 6).

8.1.2 *Probabilistically checkable proofs*

The foundational basis of STARKs is a combination of probabilistically checkable proofs, zero-knowledge proofs, and random function generation. PCPs are proofs that are large, but probabilistically checkable in a comparatively short amount of time (Arora & Safra *et al.*, 1998). Formally, probabilistic proofs are proofs in the structure of the probability of error being so small as to be extremely probabilistically true, and close enough to true when a proof might not be solvable exactly. Probabilistic proofs are used in the STARKs formulation because they are known to be unconditionally secure to any kind of attacker, quantum or classical. PCPs generate very long proofs (polynomially related to the length of the computation that is being verified). However, despite the large proof size, the verification procedure is succinct, in that the proofs do not need to be read in full, but only sampled (hence the name probabilistically checkable proofs). The verifier can check a sample large enough to give probabilistic certainty that the proof is correct.

Another foundational technology is zero-knowledge proofs, introduced by Goldwasser *et al.* (1989). A zero-knowledge proof is a proof that reveals no information except the correctness of the statement, typically in a one-bit answer. The third technology used in STARKs is random function generation. A constructive theory of randomness for functions based on computational complexity is employed (Goldreich *et al.*, 1986). Pseudorandom functions (comprising any one-way function and a random string) are produced that are indistinguishable from genuine random functions by any probabilistic polynomial-time algorithm that asks and receives the value of a function at arguments of its choice.

The three concepts are consolidated into the idea of computationally sound proofs or succinct non-interactive arguments (SNARGs), as designated in current terminology (Micali, 1994). Such computational proofs enable the verification of NP statements with much lower complexity and provide a solution to the problem of verifiably delegating computation.

The construction creates a computational zero-knowledge proof for use in the blockchain context from two elements, PCPs and cryptographic hash functions. The further advance of the contemporary zero-knowledge proof is that the prover performs and certifies the sampling process used in the proof structure, and the verifier simply confirms the proofs of the sampling results (similar to the quickly verifiable proof-of-time and proof-of-space proofs used in the IPFS/Protocol Labs system). Hash-based cryptography facilitates the entire verification process.

8.1.2.1 *Constructing a ZKP using a PCP*

Executing a zero-knowledge proof using a probabilistically checkable proof consists of four steps. These are conducting a proof, committing to the proof by instantiating it in a hash tree (a hash-linked data structure), certifiably sampling the proof with the hash-linked data structure, and compressing all of this activity into a small proof that is sent to the verifier to verify. In more detail, first, the prover conducts a PCP. Second, the prover commits to this proof in a multi-step process using a random oracle as a tool. The random oracle is a generic, agreed-upon, known method for making queries, such as a cryptographic hash function (a one-way function which is easy to verify and difficult or impossible to back-calculate).

In the second step, the prover makes a hash of each element in the proof, then a hash of each two hashes, and so on, hierarchically up until having just one top hash. This is a Merkle tree structure, and the top-level hash that contains the consolidated hashes of the whole tree is the Merkle root (the root of the Merkle tree). Bitcoin uses Merkle trees this way, rolling up the hashes of individual blocks into higher and higher levels until the top-level Merkle root is reached, which effectively calls the whole chain of all 580,000 transaction blocks that have occurred between January 2009 and June 2019. Being able to call the entire database means that the data can be verified in a matter of seconds with a one-way hash function (validating that there has not been any change to any of the data). In such a hash-validated data structure, the Merkle root can call the entire structure of digital entities such as a Github codebase, a database, a document corpus, a medical file, or a blockchain. At the highest level, one

Merkle root could call the entirety of the world's digitally-instantiated knowledge or a culture's history (a concept enumerated in science fiction as a society's data pillar (Bear, 1985)).

Third, the prover uses the Merkle root to perform and certify query samples of the proof. The Merkle root is employed to derive randomness for the query-sampling procedure so the prover is not selecting the queries directly. The prover then samples some of the queries, and establishes the authentication paths to certify the answers against the root from which the randomness was derived.

Fourth, the prover packages all of this into a very-small proof, and sends it to the verifier. Fifth, the verifier independently checks the Merkle authentication paths, confirms that the paths are certifying the answers to the queries against the Merkle root, and ultimately decides whether to accept the proof. The prover can do its own "DIY sampling" because the cryptographic hash process forces the prover to commit to the proof before the queries are sampled (otherwise the prover could cheat). In any case, the verifier confirms the queries.

The queries sample a small encoded part of the proof and assess whether it could have come from the mathematical apparatus of the proof (the hash-linked roll-up of the proof into the Merkle tree, and the random query sampling). The proof apparatus is a giant scheme for validating internal consistency. The idea is that it becomes much too difficult to cheat because of all the internal checks. Further, the apparatus allows the proof to be verified quickly. A physical-world analogy could be made to a courtroom proceeding in which a witness is asked questions related to internal consistency such as if the sun was shining that day, or if there was a traffic jam, such that eventually everything has to add up as being internally consistent in a truth claim, and after enough probabilistic rounds, inconsistencies simply cannot hold together coherently.

The same principle is used in zero-knowledge proofs, in that the method effectively introduces an apparatus for internal consistency checks. Instead of waiting hundreds of rounds in a traditional interactive proof system, the idea is to incorporate internal consistency checks into the proof structure such that the prover randomly samples, confirms, and certifies the proof before sending it to the verifier. The computational apparatus can execute many different kinds of internal consistency checks.

The prover is essentially asked to generate an encoding (with an error-correcting code) of the steps of the computation and thereby creates an artificial apparatus for the purpose of the proof that has a lot of internal consistencies that all fall into place if the statement is true, and are inconsistent if it is false. At the end, the prover sends a Merkle root of its commitment to a particular encoding of the schema to the verifier, who easily checks its validity.

The basic construction of a zero-knowledge proof using a PCP is used in STARKs. The hashing is very-fast in linear time, so most of the resource cost of the proof is in the PCP construction. The zero-knowledge proof size is small relative to the PCP size because only a small number of queries is being communicated together with the indication paths whose length is logarithmic in the length of the proof. This is different from SNARKs in which the cryptography consumes most of the computational cost (through the multi-exponentiation employed by the proof structure).

8.1.3 *PCPs of proximity and IOPs: Making PCPs more efficient*

8.1.3.1 *Probabilistically checkable proofs of proximity*

PCPs provide a working proof technology, but they are inefficient even considering the fast verification time given the lengthy proof size. PCPs are too long in length and too clunky in operation to be used in a practical setting, therefore, modifications to make them simpler and quicker are a target of research efforts. One suggestion for improved proof composition is modular proof components (Dinur & Reingold, 2004). Another suggestion for improved verification time and other aspects is PCPs of proximity (Ben-Sasson *et al.*, 2006). Probabilistically checkable proofs of proximity are a relaxation of traditional PCPs that constrain verification to nearby values so that the verifier only needs to read a smaller number of bits. PCPs of proximity more effectively manage the trade-off between the length of PCPs and their query complexity.

In a standard PCP, there is an input and a witness (a random oracle) that can be queried about the input. The verifier submits queries to the witness until becoming probabilistically satisfied that the proof is true or false. The number of queries is counted and computed into a measure of query complexity. In a PCP of proximity, to speed things up, features and constraints are added. There is both an input-oracle and a witness-oracle such that the verifier queries both. The verifier compares the query response between the input-oracle and the witness-oracle. Since the witness-oracle encodes input from the input-oracle, any queries to both should be in proximity (within a close Hamming distance) when they are compared (if the proof is true). The idea is to query a random location in the input and the corresponding location in the witness oracle, and test for proximity (Ben-Sasson *et al.*, 2006, 44). The optimization focus is on the verifier having very low query complexity, smaller than the length of the input, because this results in very-fast verification time.

8.1.4 *IOPs: Multi-round probabilistically checkable proofs*

Interactive oracle proofs (IOPs) are essentially multi-round PCPs. PCPs of proximity and other improvements constitute an advance, but are not enough to render PCPs efficient enough for practical use. More recent progress in the related area of IOPs has made PCPs feasible for implementation in the blockchain environment. Interactive *oracle* proofs are not the same as the interactive *manual* proofs connoted in the basic concept of zero-knowledge proofs. In interactive manual proofs, Alice has to request and process enough encrypted messages from Bob to become convinced that Bob is really who he says he is. The idea is to move away from manual execution by relocating the iteration cycles to automated computation. IOPs instantiate the iteration cycles in software where they can be performed extremely quickly with computation.

The reason to use interactive proofs versus non-interactive proofs is that the sequential processing allows greater complexity to be incorporated into the proof structure, yet still in a reasonable amount of execution time (in proof generation and verification). IOPs enable a complex proof to be validated quickly enough for practical use. The strategy of the proof structure being a vast apparatus (of proof, commitment, Merkle root, query sampling, and validated paths to queries) upon which internal consistency checks can be performed to detect whether the proof is true or false, can be run much more quickly with IOPs, and in a feasible amount of time.

Interactive oracle proofs are essentially a multi-round extension of probabilistic proofs. In the basic case of the regular PCP construction, the prover sends a long proof that the verifier probabilistically samples at a few locations (or that the prover randomly samples, certifies, and sends as part of the proof). In the extended interactive oracle proof version, the prover sends the verifier a few long proofs at different time intervals, which again, the verifier (or the prover) probabilistically samples at a few locations. The sequentiality introduces a dimension of time complexity and the property of ordering into the consistency apparatus of the proof structure. This quickly expands the efficiency of the proof verification, in that only a real prover knows at which time a specific action was performed, particularly relative to other actions in the proof edifice. Again, it does not matter who is doing the sampling, the prover or the verifier.

SNARKs use multi-round PCPs (i.e. interactive oracle proofs). The prover executes a long proof, commits to it with a Merkle tree, and inserts it into a hash chain to keep track of all the rounds. The new Merkle roots from the commitments to new proofs in subsequent rounds are incorporated into the hash chain (keeping track of the order with time-stamp functionality). As in non-iterative PCPs, in the query sampling phase, queries are sampled, answers obtained, and a small proof that incorporates all of the routes and the authentication paths is packaged and sent to the verifier. The verifier then checks that the answers certify against the roots (which is easy to check since the Merkle roots are in a hash chain), checks that the order of rounds was respected, and confirms that the answers are correct. The upshot is that the iterations (which perform sequence-based checks) improve the efficiency of the proof. The concept of interactive oracle proofs is a general feature which might likewise apply to other proof constructions in blockchains and beyond.

8.1.4.1 *STARKs: Reed–Solomon error-correcting codes*

The further advance of STARKs is an exponential speed-up in verification time achieved by incorporating Reed–Solomon error-correcting codes, which provide a fast (i.e. linear time) system for using error-correcting codes. Error-correcting codes are an important element in the proof structure because the prover encodes its commitment to the proof and the artificial consistency-checking apparatus through an error-correcting code. Reed-Solomon error-correcting codes allow the error-correcting code schemes to be applied quickly in STARKs.

8.1.4.2 *Deep STARKs*

A further efficiency improvement to STARKs is proposed as DEEP STARKs (Ben-Sasson *et al.*, 2019). The STARKs incorporate a mechanism called Domain Extending for Eliminating Pretenders (DEEP) which forces the prover to choose one codeword from a list of pretenders that are close to the actual answer. Regular STARKs operate by constraining the queries to sample only within a small zone in the whole field of the proof. For validity, any of these queries should be able to tell immediately if the proof is valid. However, too many queries need to be made, and so the new DEEP STARKs method allows a query to be sampled from a deeper range outside of the immediate zone and matched with a similar sample in the immediate zone.

8.1.5 *Holographic proofs and error-correcting codes*

8.1.5.1 *Holographic proofs*

Probabilistically checkable proofs are more generally part of a class of proofs known as holographic proofs, transparent proofs, or instantly checkable proofs (Babai, 1993). Holographic (transparent) proofs are called so because they have a holographic format, which renders their truth or falsity immediately transparent.

Holographic (transparent) proofs is a form of proof in which every proof or record of computation can be presented. The format has the property that the presence of errors is instantly apparent (transparent), after checking only a small fraction of the proof. Errors are calculated as essential deviations from a specified form or according to some other metric. Transparent proofs are verifiable very quickly, in time that is a constant power of the proof length. PCPs are naturally cast in the holographic form because they allow the whole of the proof to be checked with only a few samples. Holographic proofs can be used for the verification of large computations because the problem is more efficient in a holographic form. Also, the verification time grows only slightly, even if the proof size is growing exponentially.

The intuition for holographic proofs comes from holography. In a laser-generated hologram, it is possible to reconstruct the whole image from any subsection of the image. Likewise, in a holographic proof, every statement contains information about the entire proof. Therefore, an error in even a single line of the original proof has a higher probability of being visible in any given line of the holographic proof. Converting a proof into a holographic form amplifies any mistakes or falsities such that they are more easily detectable. To ascertain a proof's validity, the verifier only has to review a few statements. There is no need to examine the whole proof, irrespective of its length. Holograms operate on the principle of the part equaling the whole (a synecdoche).

8.1.5.2 *Error-correcting codes improve proof efficiency*

A holographic proof system produces proofs whose truth can be verified immediately. For efficiency, holographic proof systems are frequently implemented in the form of error-correction codes to prevent small variations in data from slowing down the algorithm. The idea is to correct small data variances or noise into a smooth form that is treated by the proof system. The claim that the proof supports is given in an error-correcting format. This is the input–output function or the matching that the proof computation is to confirm. The claim is converted into an error-correction format with whatever error-correcting code scheme is used by the proof system. Any method of error-correcting can be used, for example, the Reed-Solomon code (which corrects a constant fraction of errors in nearly linear time) is used in STARKs. The verification consists of confirming that the code is within the required distance of a unique codeword that encodes the claim supported by the proof (Levin, 1999).

The motivation for using error-correction codes in holographic proofs is to have a generic mechanism that can check computations very quickly (Babai *et al.*, 1991). One example is checking random oracle instance-witness pairs. Interactive proof protocols are used to check computations with an error-correcting code. An error-correction code is employed to transform (encode) the messages generated by the proof into unique codewords. An error-correction process can be run such that any bit of the unique codeword can be recovered (measured) within the distance area of the proof. The result is that the proof can run faster. It is possible to perform a transparent proof very quickly when data are encoded and operated on in the form of an error-correcting code scheme. Conceptually, the error correcting-code method is somewhat similar to a data smoothing function or a data compression function.

8.2 Holographic Codes

Holography is used in different ways in physics and mathematics to engage ideas related to interference, dimension-spanning (information compression), and reconstruction. A hologram is a 3D image recorded on a 2D surface. Lasers are used to create interference patterns. A hologram captures the interference pattern between the laser beams and records it on the recording medium, which is a 2D surface. Later, when the hologram is lit up in a certain way, the recorded pattern can be seen by an observer.

The dimension-spanning connotation of holograms appears in physics as the holographic principle or the holographic correspondence between a bulk region and a boundary region with one fewer dimensions (a 3D volume represented on a 2D plane). A key concept is reconstruction, the notion that a 3D volume can be reconstructed in a 2D surface, because all of the information needed to describe the 3D volume is compressed into the 2D representation. The 2D holographic image reconstructs the 3D volume. The interference-related properties of holograms are used in mathematics and computer science in the ideas of holographic algorithms and holographic proofs. Both physics and mathematics use holographic codes, which are employed in error-correction concepts and techniques, as well as other kinds of encodings.

8.2.1 *Holographic algorithms*

In computer science, holographic algorithms refer to a technique used for the efficient reduction of computational complexity (in various greater-than-NP hardness problems such as graph-theoretic (3D) problems).

One strategy in reducing computational complexity is to simplify the problem. The idea is to identify other similar problems which are easier to solve, and develop a correspondence or mapping between local solutions of one problem and local solutions of the other problem. The first reduction techniques of this nature defined correspondences as one-to-one, one-to-many, and many-to-one. However, this was limited, and a proposal was made for many-to-many correspondence based on holographic interference patterns (which are a natural form of a many-to-many correspondence).

Such holographic algorithms preserve data integrity by generating interference patterns among solution fragments to the problems in a way that conserves their sum. The main innovation is that a higher-dimensional method is produced for working with the problem. The holographic algorithms permit reductions in which the specific correspondence (or encoding) between the solution fragments of the two problems does not need to remain identifiable. Holographic algorithms are similar in structure and concept to a zero-knowledge proof, in that the idea is that it is only necessary to instantiate and work with a meta-level of data in a two-tier information system; a zero-knowledge system is only interested in whether the proof claim is true. Holographic algorithms can be seen as a technique for producing interference patterns among the solution fragments of corresponding mathematics problems, and engaging the system at this compressed level of information, hence the name holographic algorithms (Valiant, 2004, 307).

8.3 Post-quantum Cryptography: Lattices and Hash Functions

Keeping public digital infrastructure safe for the long term is a key concern. Post-quantum cryptography refers to various approaches for getting cryptography ready for the era of quantum computers. This is called post-quantum cryptography, as well as other terms such as quantum-resistant computing and post-quantum secure computing. In general, post-quantum cryptography is the study of cryptosystems which can be run on classical computers but are secure even if an adversary possesses a quantum computer. Upgrading classical computers to be quantum-ready is one focal point, and developing cryptographic algorithms to run on quantum computers is another. The idea is first keeping classical computers safe from quantum computer-based attacks, and second exploiting the capacity of quantum computers to create and run new cryptographic algorithms in the future.

Summarizing current progress, the US NIST has a robust development process for new public-key cryptography standards in sponsoring competitions and research efforts to generate algorithms for post-quantum cryptography. In January 2019, an NIST report announced that 26 algorithms were advancing to the post-quantum crypto semifinal (from 69 first-round candidates) (Alagic *et al.*, 2019). Of the 26 algorithms, 17 focus on public-key encryption and key-establishment schemes, and 9 on digital signature schemes. The approaches are mainly lattice-based, code-based, and multivariate. The lattice-based approaches most frequently target the Learning with Errors (LWE) problem, either the module or ring formulation (MLWE or RLWE). The code-based approaches mainly use error-correcting codes such as Low Density Parity Check (LDPC) codes. The multivariate approaches are based on field equations (hidden fields and small fields) and algebraic equations.

8.3.1 *Lattice-based cryptography*

In general, there are two main approaches to post-quantum cryptography, lattice-based cryptography and hash function-based cryptography. The first, lattice-based cryptography, is currently the most important candidate for post-quantum cryptography and the main method used in next-generation NIST algorithm development and beyond. Lattice-based cryptography refers to using cryptographic building blocks that involve lattices, either in proof construction or verification. Currently used public-key schemes (such as RSA, Diffie–Hellman, and elliptic curve cryptosystems) are thought to be vulnerable to a quantum computer. Some lattice-based constructions appear to be resistant to attack by both classical and quantum computers, and it is known that certain well-studied computational lattice problems cannot be solved efficiently. Lattice-based cryptosystems are based on the known hardness of lattice problems. Further, lattice problems have strong security proofs based on worst-case hardness (not merely average-case hardness as current cryptography standards).

Cryptography is based on mathematics problems that are known to be difficult to solve. The currently used cryptographic schemes are produced from a branch of mathematics called number theory, whereas lattice-based methods are derived from other branches called group theory and order theory. Very basically, number theory concerns the relation of numbers in operations such as factoring, whereas group and order theory, and specifically lattices, focus on complicated 3D problems such as the arrangement of atoms to form crystals and other materials.

8.3.2 *What is a lattice?*

The word lattice has different meanings in different contexts, but the central idea is of a regular geometric arrangement of points in a space. An example of an everyday lattice is a grid-shaped garden fence made with a framework of crossed wooden strips. The same lattice concept is used in materials science to denote the structure of crystals, solid materials whose constituent atoms are arranged in a regular 3D order. In mathematics, lattices are used as abstract objects in group theory, order theory, and set theory. Relevant for cryptography is the idea of lattices over vector spaces in group theory.

In the cryptography context, a lattice is a set of points in an n-dimensional space with a periodic structure (Micciancio & Regev, 2009). The main problem formulations in lattice-based cryptography involve calculating optimization problems given a sequence of vectors in the n-dimensional space. The basic problem is the shortest vector problem (SVP). In the SVP, the basis (the center of the lattice) is arbitrary, and the objective is to find the shortest non-zero vector in the lattice space given other input parameters.

Inaugurating the field of lattice-based cryptography is a proposal for a probabilistic public-key cryptosystem that is argued to be quantum-secure, assuming that the SVP (a very-difficult problem) cannot be solved in polynomial time (Ajtai & Dwork, 1997). The shortest vector problem is framed as "finding the shortest non-zero vector in an n-dimensional lattice, in which the shortest vector v is unique, in the sense that any other vector whose length is at most a constant power of n times the length of v, is parallel to v" (*Ibid.*, 284). The technical implication is that lattice-based cryptosystems that are based on the shortest vector problem are likely to be quantum-secure, unless the worst-case lattice problem can be solved in polynomial time.

8.3.2.1 *Lattice-based cryptography: Current research topics*

The field of lattice-based cryptography is becoming more robust. The main research theme is finding lattice-based analogs for traditional cryptography formulations. This includes defining problem classes and methods, determining what is solvable and not solvable, considering non-interactive and interactive proofs, and developing better lattice-based proof systems. Important contemporary advances are in the areas of zero-knowledge proof systems, especially in the blockchain context.

8.3.3 *Lattice-based cryptography and zero-knowledge proofs*

Some important early work proposes non-interactive zero-knowledge proof systems for lattice problems (Peikert & Vaikuntanathan, 2008). The model constructs non-interactive statistical zero-knowledge proof systems for a variety of standard approximation problems on lattices such as the shortest independent vector problem. Before this, proof systems for lattice problems were either interactive or leaked knowledge (or both). These are the first non-interactive statistical zero-knowledge proof systems that target problems not related to integer factorization, but rather lattice-based shortest vectors. Recent research in this lineage includes lattice-based non-interactive zero-knowledge proof systems for NP-hard problems (Kim & Wu, 2018) and NP-hard problems based on the LWE problem (a worst-case lattice problem) (Peikert & Shiehian, 2019).

8.3.4 *Lattice-based cryptography and blockchains*

Succinct non-interactive arguments (SNARGs) enable the verification of NP computations with substantially lower complexity than is required for classical NP problem verification (Micali, 1994). However, SNARGs and SNARKs (blockchain implementations of SNARGs) rely on pre-quantum assumptions and are not post-quantum secure. Hence, lattice-based SNARGs are proposed (Boneh *et al.*, 2017). This lattice-based SNARG method uses pairings-based cryptography (the pairing of elements between two cryptographic groups). The formulation is extended in a practical construction of SNARKs using fully lattice-based building blocks instead of pairings-based cryptography (Menon, 2017). Other work proposes a lattice-based SNARK protocol in which a proof consists of five LWE encodings based on a technique called Square Span Programs (SSPs) (Gennaro *et al.*, 2019).

Lattice-based SNARGs and SNARKs could be one route to post-quantum cryptography for blockchains, however, the proposed solutions thus far require a trusted verifier, so they are not fully self-contained or decentralized. STARKs, and hash-function cryptography more generally, is another route to post-quantum cryptography for blockchains. SNARKs are conjectured to be fully post-quantum because they rely on hash functions, and do not require a trusted verifier.

8.3.5 *Hash function-based cryptography*

Aside from lattice-based cryptography as the primary method in development for post-quantum cryptography, the other main approach is hash function-based cryptography. Claims about what kinds of cryptography may be post-quantum or not can be confusing. On the one hand, a key claim is that constructions that only make use of hash-based functions are post-quantum. On the other hand, the idea that most readily comes to mind regarding quantum computing is that it may be able to break current encryption standards which are based on hash functions. How can both be true?

A clarification can be made that hash functions are a generic technology that is widely used for different purposes. Any particular hash function might be quantum-secure or not depending on the underlying technology it uses to create the hashes. (A hash function is a software program for creating a short digest or representation (say, always 32 characters or 64 characters) of a longer data file.)

The point is that the underlying technology used to create the hashes in standard public-key cryptography at present (2048-bit RSA encryption) is not quantum-secure. Potentially quantum-breakable technology includes hash functions based on 2048-bit RSA encryption, elliptical curve cryptography (used for digital signatures in blockchains), and any other kind of cryptography based on factoring and discrete logarithms, pairings, and knowledge assumptions (such as the knowledge-of-exponent assumption).

Hash functions based on other technologies (that are not quantum-breakable algorithms) may be quantum-secure. For example, a lattice-based hash function might be quantum-secure. STARKs, which hash the output of PCPs, are conjectured to be quantum-secure. A hash function itself is a generic structure that can be upgradable to post-quantum technologies as they become standard.

8.4 Quantum Proofs

In quantum proofs, the prover and the verifier exchange and process quantum information instead of classical information. In computational complexity theory, a proof is a computation (a mechanical process). The traditional notion of a proof in mathematics consists of an individual who proves a theorem (the prover) and another individual who verifies the proof (the prover). Computer-assisted or computational proofs quickly became an expedient way to conduct proofs. Since they are computations, proofs can be analyzed with complexity theory. Computational complexity is calculated for the two parts of the proof, assessing both efficient construction and efficient verification. A key principle (in both classical and quantum proofs) is that it takes longer to generate the proof than to check it. In both classical and quantum settings, proof generation takes NP time (non-polynomial time) and proof verification takes polynomial time. Notably, the verification takes place in one fewer time dimensions. In a mixed computation setting, it might take an equal amount of time to generate a proof in a quantum system and verify it in classical system. This suggests time arbitrage between computational complexity domains as a design principle.

There are extremely interesting implications for proofs and computational complexity theory in quantum computing. For example, in non-interactive proofs, the quantum computational analog of the complexity class NP is known as QMA. In QMA, a quantum state could play the role of the proof (also called the certificate or witness), by generating the proof in NP time, and having it verified in a polynomial time quantum computation. The computational complexity classes become the prover and the verifier. Being able to think in terms of different time regimes, and the arbitrage possibilities between them and related security implications becomes even more important. The fact that a quantum proof state could be a superposition over exponentially many classical states may offer a substantial computational advantage over classical proofs. In interactive proofs, various new quantum complexity classes arise such as QIP, QSZK, and QMIP as the quantum analogs of IP, SZK, and MIP (Vidick & Watrous, 2015).

8.4.1 *Non-interactive and interactive proofs*

There are different proof formats, namely non-interactive and interactive. One is not better than the other, rather there are trade-offs in using the different variations. The key difference is that interactive proofs are used when the verifier wants to make use of randomness in the verification process such as in the STARKs example. A non-interactive proof is a proof that does not require an interaction to take place between the prover and the verifier. The proof generation is a computational operation which is abstracted as strings of symbols and verified by another computational operation. An interactive proof (in computational complexity theory) is an abstract machine that models computation as the exchange of messages between two parties.

Whether non-interactive or interactive, in computational complexity theory, a proof is a computational process. This is distinct from a proof in real-life such as having to present a passport, or the zero-knowledge proof example of Bob sending encrypted messages to Alice until he is convinced that it is really her. In non-interactive proofs, one computational process generates the proof and another process confirms it. In interactive proofs, the computational process models an interaction between the prover and the verifier, introducing randomness to make the proof better and quicker. Either takes place very quickly in the time frame of computation, "interactive" does not mean having to wait for a human party to do something. Interactive proofs may involve both single-prover systems and multi-prover systems.

The computational complexity class related to zero-knowledge proofs is statistical zero-knowledge (SZK). SZK refers to proof results that may not be perfectly precise, but are close enough to be statistically valid (Kobayashi, 2003). The proof condition that the verifier learns nothing is interpreted in a strong statistical sense. The quantum analog to statistical zero-knowledge is quantum statistical zero-knowledge (QSZK) (Watrous, 2002). QSZK is defined as the set of computational problems with yes–no answers for which the prover can always convince the verifier of yes instances, but will fail with high probability for no instances.

It is known that the quantum model is at least as powerful as the classical one, and that all SZK problems are contained in the QSZK class. QSZK also trivially contains BQP, the class of all problems that can be solved on a quantum computer. BQP (Bounded error, Quantum, Polynomial time) denotes the class of problems that are solvable by a quantum computer in polynomial time (Bernstein & Vazirani, 1997).

The salient point of the BQP class is that for such problems, the verification can be performed by the computer itself without any need for interactive discussions with a prover. QSZK should therefore be understood as the set of problems whose yes instances can be reliably identified using a quantum computer (Watrous, 2009). To assert that all the problems in QSZK can be solved in quantum polynomial time, that QSZK = BQP, is to assert that the prover is ultimately no help at all. Another research in quantum interactive proof systems has a somewhat similar finding of superfluous elements, in that messages between the prover and the verifier can be eliminated without changing the power of the model (Beigi *et al.*, 2011).

In the classical case, there is an efficiency benefit to introducing randomness. In the quantum context, these kinds of advantages might be exploited to an even greater degree considering the quantum mechanical formulation of randomness as entropy. For example, could there be a proof technology that calculates in area instead of volume using the quantum entanglement area entropy law (thereby arbitraging computational complexity classes). Proofs that are vastly more complex yet still time-efficient might be available in the quantum domain.

8.4.2 *Conclusion on quantum proofs*

Overall, the lens of computational complexity applied in the quantum computing domain to quantum proofs indicates two important points. First is the notion of diverse time regimes, between quantum computing and classical computing, and proof generation and proof verification. Verification occurs in one-fewer dimensions than generation. One-way hash functions likewise run in one-fewer dimensions than their production too are produced in occur is a similar example. Second is the implication that the prover is unnecessary in advanced proof technologies since the technology automatically performs this function.

8.5 Post-quantum Random Oracle Model

The quantum random oracle model is one of the single most challenging risks to quantum cryptographic security (Boneh *et al.*, 2011). This is because it indicates a greater acceleration of the advantage of quantum computing over classical computing than other comparisons. The quantum random oracle model is a quantum-accessible random oracle. The adversary can query the random oracle with quantum states and evaluate the oracle in superposition. The attacker has access to a known random oracle structure (some random function such as a hash function) which it can evaluate in superposition.

A random oracle is an oracle (a generic technology) that responds to every unique query with a random response chosen uniformly from its output domain. Random oracles are used in proof systems to check if a sample is within the random oracle output range; if yes, suggesting a valid proof, if no, an invalid proof. The random oracle in computational models typically uses an abstraction of a known cryptographic hash function, currently SHA-256, Keccak, or Blake. A known cryptographic hash function is used for two reasons. First, it is a proven and established standard (so it is assumed to work), and second, it serves as an independent party to query to obtain valid quantum statistical query distributions for the proof.

Since the random oracle is known and widely implemented, it makes sense to assume that a quantum attacker can access the random oracle, and evaluate it in superposition. The random oracle is particularly threatening because of all of its potential random states, which exponentiates the potential advantage of a quantum attacker over a classical system.

8.6 Quantum Cryptography Futures

The long-term goal is not just to quantum-protect existing cryptographic methods, but eventually to quantum-instantiate new kinds of cryptographic systems. A fully-quantum proof ecosystem might be envisaged with quantum key distribution, quantum algorithms, quantum proofs, and quantum accumulators, all running on quantum computing infrastructure. Quantum accumulators are the post-quantum analog to Merkle trees (accumulating paths to system information). Different kinds of post-quantum accumulators are proposed such as RSA-style accumulators and class group-style accumulators (Boneh *et al.*, 2018). Post-quantum zero-knowledge proofs for accumulators are also proposed (Derler *et al.*, 2018), as every part of the infrastructure demonstrates its integrity with cryptographic proofs.

8.6.1 *Non-Euclidean lattice-based cryptosystems*

One direction of advance in post-quantum cryptography could be integrating the mathematical and quantum mechanical concepts and methods of lattices more profoundly. Lattices in mathematics are treated primarily with Euclidean geometry, but quantum mechanically can be applied to different-shaped geometries, such as spherical and hyperbolic geometries, and Riemann elliptical geometries. Lattices are a mathematical object, but are also fundamental quantum mechanical structures. This implies all of the interesting properties of quantum mechanical systems such as n-dimensional Hilbert spaces, superposition, entanglement, interference, and error correction.

Since mathematical lattice geometry is regular, there could be an opportunity to develop proof systems based on the alternative geometries of non-Euclidean lattices. The intuition is that the greater complexity of the environment suggests more complex proof structures that are still time-efficient to verify (essentially the principle of STARKs). Some of the non-Euclidean aspects of lattice models that are being explored in condensed matter physics and quantum computing likewise might be applied to non-Euclidean lattice cryptography. For instance, in quantum computing, quantum mechanical lattices are used in the topological matter approach (braided fermion paths) with zero modes and unique non-Abelian braiding statistics that might be suitable for cryptographic operations. Quantum mechanical lattices also feature in the examination of lattice potentials and fermions in holographic (dual representation) non-Fermi-liquids (Liu *et al.*, 2012), and the holographic reconstruction of bulk geometry from lattice simulations (Rinaldi *et al.*, 2018).

These kinds of models might be used to develop non-Euclidean instantiations of quantum mechanical lattices for non-Euclidean lattice-based cryptography. The technophysics analytic tool for this is tensor networks, and in particular, using random tensors to produce and verify the non-Euclidean lattice-based encryption and proofs.

References

Ajtai, M. & Dwork, C. (1997). A public-key cryptosystem with worst-case/average-case equivalence. In: *Proceedings of the 29th Annual ACM Symposium on the Theory of Computing*, pp. 284–93.

Alagic, G., Alperin-Sheriff, J., Apon, D. *et al.* (2019). Status Report on the First Round of the NIST Post-Quantum Cryptography Standardization Process. NISTIR 8240.

Arora, S. & Safra, S. (1998). Probabilistic checking of proofs: A new characterization of NP. *JACM* 45(1):70–122.

Babai, L. (1993). Transparent (holographic) proofs. In: Enjalbert, P., Finkel, A. & Wagner, K.W. (eds). *STACS 93. Lecture Notes in Computer Science*, Vol. 665. Berlin: Springer.

Babai, L., Fortnow, L., Levin, L.A. & Szegedy, M. (1991). Checking computations in polylogarithmic time. In: *Proceedings: 23rd Annual ACM Symposium on Theory of Computing*. New Orleans, LA. May 6–8, pp. 21–31.

Bear, G. (1985). *Eon*. New York, NY: Tor Books.

Beigi, S., Shor, P. & Watrous, J. (2011). Quantum interactive proofs with short messages. *Theory Comput.* 7:101–17.

Ben-Sasson, E., Bentov, I., Horesh, Y. & Riabzev, M. (2017). Fast Reed-Solomon interactive oracle proofs of proximity (2nd revision). *Electronic Colloquium on Computational Complexity* (*ECCC*). 24(134).

Ben-Sasson, E., Bentov, I., Horesh, Y. & Riabzev, M. (2018). Scalable, transparent, and post-quantum secure computational integrity. ia.cr/2018/046.

Ben-Sasson, E., Chiesa, A., Garman, C. *et al.* (2014). Zerocash: Decentralized anonymous payments from Bitcoin. In: *Proceedings of the IEEE Symposium on Security & Privacy*. Oakland, CA, pp. 459–74.

Ben-Sasson, E., Goldberg, L., Kopparty, S. & Saraf, S. (2019). DEEP-FRI: Sampling outside the box improves soundness. ia.cr/2019/336.

Ben-Sasson, E., Goldreich, O., Harsha, P. *et al.* (2006). Robust PCPs of proximity, shorter PCPs, and applications to coding. *SIAM J. Comput.* 36(4): 889–974.

Bernstein, E. & Vazirani, U. (1997). Quantum complexity theory. *SIAM J. Comput.* 26(5):1411–73.

Boneh, D., Bunz, B. & Fisch, B. (2018). Batching techniques for accumulators with applications to IOPs and stateless blockchains. ia.cr/2018/1188.

Boneh, D., Dagdelen, O. & Fischlin, M. (2011). Random oracles in a quantum world. *Advances in Cryptology. ASIACRYPT 2011*. 41–69.

Boneh, D., Ishai, Y., Sahai, A. & Wu, D.J. (2017). Lattice-based SNARGs and their application to more efficient obfuscation. In: Coron, J.S. & Nielsen, J. (eds). *Advances in Cryptology. EUROCRYPT 2017. Lecture Notes in Computer Science*. Vol. 10212. Cham, Switzerland: Springer, pp. 247–77.

Bunz, B., Bootle, J., Boneh, D. *et al.* (2018). Bulletproofs: Short proofs for confidential transactions and more. In: *39th IEEE Symposium on Security and Privacy 2018*.

Derler, D., Ramacher, S. & Slamanig, D. (2018). Post-quantum zero-knowledge proofs for accumulators with applications to ring signatures from symmetric-key primitives. In: Lange, T. & Steinwandt, R. (eds). *Post-Quantum Cryptography*. London: Springer.

Dinur, I. & Reingold, O. (2004). Assignment testers: Towards a combinatorial proof of the PCP theorem. In: *Proceedings: 45th Annual IEEE Symposium on Foundations of Computer Science, FOCS '04*, pp. 155–64.

Gennaro, R., Minelli, M., Nitulescu, A. & Orr, M. (2019). Lattice-based zk-SNARKs from square span programs. In: *CCS '18 Proceedings of the 2018 ACM SIGSAC Conference on Computer and Communications Security*. Toronto, Canada. October 15–19, pp. 556–73.

Goldreich, O., Goldwasser, S. & Micali, S. (1986). How to construct random functions. *JACM* 33(4):792–807.

Goldwasser, S., Micali, S. & Rackoff, C. (1989). The knowledge complexity of interactive proof systems. *SIAM J. Comput.* 18(1):186–208.

Kim, S. & Wu, D.J. (2018). Multi-theorem preprocessing NIZKs from lattices. ia.cr/2018/272.

Kobayashi, H. (2003). Non-interactive quantum perfect and statistical zero-knowledge. In: Ibaraki T., Katoh N. & Ono H. (eds). *Algorithms and Computation. ISAAC 2003. Lecture Notes in Computer Science*. Vol. 2906. Berlin: Springer, pp. 178–88.

Levin, L.A. (1999). Holographic proofs. *Encyclopedia of Mathematics*. Dordrecht, NL: Kluwer Academic Publishers.

Liu, Y., Schalm, K., Sun, Y.W. & Zaanen, J. (2012). Lattice potentials and fermions in holographic non Fermi-liquids: hybridizing local quantum criticality. *J. High Energ. Phys*. 36.

Menon, S. (2017). Implementing lattice-based cryptography in libsnark. crypto.stanford.edu. Accessed June 30, 2019.

Micali, S. (1994). CS proofs. In: *35th Annual Symposium on Foundations of Computer Science*. Santa Fe, New Mexico, November 20–22, pp. 436–53.

Micciancio, D. & Regev, O. (2009). Lattice-based cryptography. In: Bernstein D.J., Buchmann J. & Dahmen E. (eds). *Post-Quantum Cryptography*. Berlin, Germany: Springer.

Peikert, C. & Shiehian, S. (2019). Noninteractive zero knowledge for NP from (plain) learning with errors. ia.cr/2019/158.

Peikert, C. & Vaikuntanathan, V. (2008). Noninteractive statistical zero-knowledge proofs for lattice problems. In: Wagner D. (ed.), *Advances in Cryptology. CRYPTO 2008. Lecture Notes in Computer Science*. Vol. 5157. Berlin, Germany: Springer.

Rinaldi, E., Berkowitz, E. & Hanada, M. (2018). Toward holographic reconstruction of bulk geometry from lattice simulations. *J. High Energ. Phys*. 42.

Valiant, L. (2004). Holographic algorithms. In: *FOCS 2004*. Rome, Italy: IEEE Computer Society, October 17–19, pp. 306–15.

Vidick, T. & Watrous, J. (2015). Quantum proofs. *Foundations and Trends in Theor. Comput. Sci*. 11(1–2):1–215.

Watrous, J. (2002). Quantum statistical zero-knowledge. arXiv:quant-ph/0202111.

Watrous, J. (2009). Zero-knowledge against quantum attacks. *SIAM J. Comput.* 39(1):25–58.

Part 3

Machine Learning and Artificial Intelligence

Chapter 9

Classical Machine Learning

Abstract

Machine learning refers to mechanistic systems that "learn" by modeling high-level abstractions in data and cycling through trial-and-error guesses with feedback to establish an optimally weighted system that can make accurate predictions about new data. The main forms of deep learning networks are convolutional neural nets (CNNs) for image recognition and recurrent neural nets (RNNs) for text and speech recognition. Learning may be supervised (executed on pre-classified datasets) or unsupervised (finding patterns in unlabeled data). Prominent advances include adversarial nets, dark knowledge, and manifold learning. The technical aspects of machine learning include formulating the problem as a logistic regression, running it on a modular processing network, and iterating to find an optimal solution. Although machine learning excels at certain tasks, namely image and text recognition, its expandability to more complicated data analysis problems has been questioned.

9.1 Machine Learning and Deep Learning Neural Networks

Artificial intelligence, using computers to do cognitive work (physical or mental) that usually requires a human, raises issues that range from the benefits of labor-saving automation to technological unemployment and the nature of human identity (which has often been defined by work)

(Swan, 2017). In academic study, artificial intelligence is a sub-field of computer science. Within artificial intelligence, machine learning is one of the biggest focal areas. Machine learning is the study of algorithms and statistical models used by computers to perform tasks by relying on information patterns and inference as opposed to explicit instructions. Machine learning has outperformed conventional methods (digital and human) in various fields with demonstrated success in image recognition, machine translation, and speech recognition.

A widely used technique for developing machine learning algorithms is neural networks, a computer system architecture that is modeled on the human brain and nervous system. Neural networks comprise multiple layers of computational processing to instantiate successively higher levels of abstraction. Computational neural networks attempt to imitate the way the brain extracts multiple levels of representations from sensory input and has signaling threshold levels that trigger action. Deep learning neural networks are the latest incarnation of these methods. So-called "deep networks" were proposed as multi-layer neural networks with top-down connections (Hinton, 2007), and the term "deep learning" arose to describe Google's successful cat and facial image recognition project based on YouTube data (Le *et al.*, 2012).

Conceptually, deep learning is a computer software program that can identify what an object is, either a physical object or a digital object. More technically, deep learning is a class of machine learning algorithms in the form of a neural network that uses a cascade of layers of processing units to model high-level abstractions in data and extract features with which to make predictive guesses about new data. Mathematically, deep learning is the instantiation of a problem with logistic regression, applied to a computation graph of modular processing units with variable weights and biases, to create a mapping function between input data and output guesses, which is optimized by cycling forward and backward through the layers using a loss function (such as gradient descent) to obtain predictive results. A representative task for a deep learning network, in autonomous driving for example, is to identify whether another object in the landscape is a car or not. The immediate goal of deep learning networks is to create an image, speech, or text recognition system that determines which features are relevant (at increasingly higher levels of abstraction) from a set of training data such that new examples can be correctly identified.

9.1.1 *Why is deep learning called "deep"?*

Deep learning is called deep in the sense of there being "hidden" layers in the network architecture. The deep learning system is constructed like a sandwich, with an input and output layer at each end, and processing layers between them whose operation is hidden to the user. The system uses the hidden layers of computational processing to map functions between the input data and output guesses. Whereas shallow networks have only one or two layers of processing, deep networks have three and possibly many more layers.

A basic demonstration case for contemporary deep learning networks is 5–8 layers. On the larger size, GoogleNet's 22-layer network for image classification and object recognition is one of the largest existing networks. In research, the theoretical possibilities are higher, suggesting that with stochastic depth (randomly determined layers), networks might be increased in the order of 1200 layers and still yield meaningful results with error-decreasing improvements (Huang *et al.*, 2016). Optimal network structure is a complex problem in which depth (number of layers) is only one factor.

9.1.2 *Why is deep learning called "learning"?*

Deep learning is called learning in the sense that the system "learns" by creating a mapping function from input to output to identify salient features in a dataset for the purpose of accurately identifying new examples. Deep learning networks are mechanistic systems that learn by cycling through trial-and-error guesses with feedback to establish higher levels of abstraction and probabilistic solutions. A deep learning system is "dumb" in the sense that it is a mechanical system like a Searle's Chinese Room (simply executing tasks without actually "understanding" what it is doing). As any computer, the deep learning system simply executes algorithmically specified instructions. In this sense, deep learning is a programmed optimization method for discovering the best set of parameters that correctly identify new examples of the kinds of items that it has been trained to classify.

Deep learning is based on the principles of the artificial intelligence argument. The artificial intelligence argument is that basic data classification problems may be solved by running fairly straightforward algorithms over very-large datasets. Having large enough data is what makes the difference. Indeed, this method has proven successful so far in artificial intelligence advances, in contrast to other methods such as creating expert systems and attempting to digitally catalog the entirety of human knowledge. Large data corpora have enabled progress in machine translation (Halevy *et al.*, 2009), the image recognition of cats and faces (Le *et al.*, 2012), and systems with greater than human-level capability in object recognition (He *et al.*, 2015).

9.1.3 *Big data is not smart data*

The reason that data analysis techniques such as deep learning are needed is to keep pace with the exponential rate of data growth. The modern era is one of big data, characterized by very large datasets that are too big and complex for traditional data processing methods. Big data is characterized by three "Vs": volume (size), velocity (speed), and variety (text, graphics, video, and audio data), and comprises very large datasets that must be analyzed computationally to find patterns, trends, and associations. IDC estimates that global data volumes will reach 40 zettabytes in 2020, of which only 1% are analyzed and 20% are protected (Burn-Murdoch, 2012). The problem is that the rate of data growth eclipses the ability to process it, and consequently, *big data* is not *smart data*. Only 42% of organizations say that they are able to extract meaningful insights from the data available to them (*Oxford Economics*, 2018), even though data analytics and artificial intelligence initiatives are crucial to strategic plans for digital transformation. Conventional statistical analysis techniques such as regression work well on small data sets, but large datasets require different and more sophisticated tools (Varian, 2014). As a result, *data science* has arisen as an academic and vocational field for the application of scientific methods for data analysis (Cleveland, 2001).

9.1.3.1 *Size of deep learning networks*

AlexNet (2012) is an early example of an image recognition network, using eight layers, 60 million parameters (variables that determine the network structure), and 650,000 processing nodes to classify 1.3 million images (Krizhevsky *et al.*, 2012). In the 2012 ImageNet LSVRC competition, AlexNet beat other networks, performing at an error rate of 15.3% versus 26.2% (second place) (Gao, 2017). Since then, deep learning networks have grown. Microsoft's ResNet launched in 2015 with 7 exaFLOPS of processing power and 60 million parameters. Baidu's Deep Speech 2 model was introduced in 2016 with 20 exaFLOPs and 300 million parameters. In 2017, Google's Neural Machine Translation system had 105 exaFLOPS and 8.7 billion parameters (Sherbin, 2017). Digital Reasoning, a government contractor based in Nashville TN that provides cognitive computing services, has a much larger system with 160 billion parameters (Trask, 2018). Deep learning networks are run on *training data* sets overnight or for a few days, and then the resulting algorithms are applied to new sets of *test data*. Deep learning performance is reported as the error rate of the network, meaning the percent of data not classified correctly.

9.1.4 *Types of deep learning networks*

9.1.4.1 *Supervised and unsupervised learning*

One distinction in deep learning networks is between supervised and unsupervised learning. Supervised learning is used when data sets are already classified and labeled. Unsupervised learning is used to detect patterns in unlabeled data. In supervised learning, there is an explicit training phase, applying techniques such as regression analysis to train a learning system, with data already labeled as input and output pairs, to develop predictions about new test data. In unsupervised learning, there is no explicit training phase and the learning system investigates potential patterns in the input data directly. Typical unsupervised learning tasks include cluster analysis and feature extraction. Cluster analysis is dividing data into different groups based on certain measures of similarity. Feature extraction is identifying salient properties that can be used to describe the data.

Unsupervised learning may be used as a pre-processing step to characterize data before supervised learning is applied. For example, in condensed matter physics, unsupervised learning may be applied to assess the overall characteristics of the system such as phase transitions (Wang, 2016), whereas supervised learning is applied to study the specific phases of matter (Carrasquilla & Melko, 2016).

9.1.4.2 *Convolutional (image) and recurrent (text, speech) nets*

Aside from supervised and unsupervised learning, another primary distinction in deep neural networks is between convolutional neural nets (CNNs) used for image recognition and recurrent neural nets (RNNs) used for text, speech, and audio recognition. CNNs convolve, which in this context means progressively rolling up to higher levels of abstraction in feature sets (by analogy to the mathematical function of convolving by combining or blending one function with another).

RNNs recur (iterate), analyzing data in which sequence matters such as natural language and time series data. Having some sort of memory function to identify and recall relevant sequences is necessary, and the long–short-term memory (LSTM) mechanism is often used. LSTM has intelligence built in to figure out what to forget, and what to store for the short-term and longer-term processing phases of the network. Memory is costly, so it must be efficient. An example is the network detecting that the salient element of the sequence to store in the cycle of someone making fish, chicken, and beef for dinner on consecutive nights might be what was made yesterday, as opposed to which day of the week it is.

CNNs and RNNs are the two primary architectures for deep networks, but there are others, for example, generative adversarial networks and reinforcement networks, which are recent advances. In adversarial networks, there are two networks. The adversary network generates false data and the discriminator network tries to learn whether the data are false (Creswell *et al.*, 2018). Reinforcement networks have a goal-oriented algorithm in which the system learns to attain a complex objective over many steps.

9.2 Perceptron Processing Units

Deep learning networks comprise thousands or millions of modular processing nodes, which are identical and assembled into layers in a Lego-like fashion. The architecture is highly scalable and networks can be organized into arbitrarily many layers. The processing units may be variously called perceptrons, artificial neurons, neurons, logits, or nodes, all of which are (generally) synonymous. The structure of the processing node is derived from graph theory, with edges (lines indicating input and output values) and nodes (centers where processing takes place).

The node computation is in the form of a computation graph with three phases: input, processing, and output. The incoming edges carry input data values (numbers) that come into the node to be processed. The node performs an operation (such as addition) on the input data values. This results in an output data value (a number) that becomes one of the input data values for the downstream node in the next layer. The operations conducted by the node are simple pre-specified calculations. The basic computation of the node structure is fixed, but there are other variables which the network adjusts in its learning process.

Each node has two variable parameters, a weight and a bias. The input values may be weighted (applying the weight as a coefficient), and the node processing center may have a bias (applying the bias as a threshold value). The weights and the biases are the variable parameters that the system adjusts with trial-and-error guessing as it tries to identify relevant features in the data set. The input and output values change as the network operates. The weights assigned to the input edges vary as learning proceeds, increasing or decreasing the strength of the signal that is sent into the node to be processed. Nodes may have a threshold value (bias) such that the downstream signal is sent only if the aggregate signal is above or below a certain level (analogous to signal transmission thresholds in the brain). The system cycles forward and backward (feeding forward and back-propagating), iteratively updating the variables (weights and biases) as it establishes which features are most relevant for making correct guesses about the training data. A deep learning system is a vast modular network of layers of processing nodes, with variable information flows cascading forward and backward across the network to find optimal answers.

9.2.1 *Jaw line or square of color is a relevant feature?*

The way the system learns from input data is first by identifying and modeling all possible features, and then by adjusting the probabilities assigned to specific features to test the relevancy for identifying the general type of the item in question. For example, a deep learning system initially catalogs all possible features in an image, not knowing which will be important. This could include the jaw line of a face and a square of blue color from a T-shirt. Then, in training on multiple images of faces, the system learns which features recur and are therefore relevant. The "jaw line" feature persists in multiple images but the "square of blue cloth" feature does not.

The recurrence of features is instantiated as a feedback loop that the system uses to vary the weight and bias variables, such that it can learn to correctly identify faces. The feature weights may start as being equally distributed (0.5 on each input edge), and are then adjusted (say) to 0.75 for an upweighted feature (jaw line) and 0.25 for a down-weighted feature (blue cloth). This is how the system mechanically identifies the relevant features in facial recognition, and scales up the model with lots of examples of the "right answer" (images of faces), such that it can make predictive guesses about new data (whether a new image presented to the system is likely to be a face). The statistical basis of learning systems is seen as the statistically frequent features quickly surface and the system uses probability to guess correct classifications.

9.3 Technical Principles of Deep Learning Networks

There are three technical (mathematical) aspects of the operation of deep learning networks: logistic regression (statistics), matrix multiplication (linear algebra), and optimization (calculus). First, the problem is formulated as a logistic regression. Second, the logistic regression is instantiated using a modular network of processing units organized into cascading layers and processed with matrix multiplication. Third, an optimal answer to the logistic regression is obtained by cycling forward and backward through the network, altering the weight and bias variables with a loss function.

9.3.1 *Logistic regression: s-curve functions*

The first step in operating deep networks is defining the problem in terms of a logistic regression. Regression is a standard data analysis technique used to find an equation or function that describes a set of data. A function is a higher-order representation of the same information. For example, different levels of representation of the same data can be specified as the plotted data, a linear regression, and a logistic regression. Each subsequent level represents the data more abstractly. Linear regression is useful for predicting a continuous set of values such as house price per house size. However, as the term suggests, linear regression is limited to situations in which there is a linear relationship between the dependent and independent variables (in the house example, usually the larger the house, the higher the price).

Logistic regression on the other hand, allows the same kind of discrete outcome to be predicted, but from a larger set of more complex input variables. Input data may have different forms and relationships (e.g. continuous, discrete, dichotomous, or mixed). Logistic regression can be used to predict a binomial value (Yes/No, 1/0), or a probabilistic value (between 0 and 1, 0–100%) for a dependent variable, based on one or more independent variables. For example, the problem might be to predict the probability as to whether a certain individual will become a company's customer based on multiple independent variables such as income, education, mortgage, and work experience. Using logistic regression, these variables can be mapped into one higher-order equation in which the outcome is interpreted as the probability of any particular individual becoming a customer.

The logistic regression is represented as an s-shaped curve that compresses an entire range of values into an easily readable nonlinear probability function. The function accommodates different output values, all in the same structure, whether binary values of Yes/No, or 1/0 (in which any value between 0.5 and 1 is read as Yes), probability values between 0 and 1, or hyperbolic tangent (Tanh) values between (–1) and 1. The output values can be readily understood (by human or machine), and easily compared with other values (a 73% feature is stronger than a 57% feature, for example). A further benefit of the s-curve formulation is that optimization tools from calculus may be applied, such as calculating the minimum and maximum points of the curve.

9.3.2 *Modular processing network node structure*

The second aspect of the operation of deep learning systems is the modular architecture of the network. The central feature is the processing node, which executes the computational phases related to input, processing, and output. Through trial-and-error adjustment of the weights and bias variables of the processing nodes, the objective of the deep learning system is to generate a function that successfully maps input data to output guesses such that new data can be correctly identified.

9.3.2.1 *Digitizing input data into an array*

Considering the example of an image recognition engine, the input data are images. The MNIST training data, a 60,000-item database of handwritten digits, is often the standard dataset employed to test deep learning systems. In a simple example, images of 0s and 1s are used. Each image is digitized, meaning converted into numbers that the system can use to perform computations. Each image is divided into a grid (a 28 × 28 pixel grid, for example), and a value is assigned to each pixel (square in the grid) based on its brightness (a value of 0–255) or some other variable. The original image is represented as a grid of brightness values. If the numerals are white on a black background, there will be a 0-brightness value (black) for most pixels, and a 255-brightness value (white) for the pixels that correspond to the numeral portion of the image (and some intermediary brightness values around the numeral borders).

The brightness values are read into an array (a long row of comma-separated values, like one long row in an Excel spreadsheet). A 28 × 28 pixel grid translates into an array or string of numbers that has 784 elements or brightness values, separated by commas. Each of the input images is mapped into its own spreadsheet row with 784 cells, each cell corresponding to the brightness of a certain location in the 28 × 28 grid. Digitized into arrays in this manner, the input data (784-elements per image) is fed into the deep learning network. The arrays corresponding to the images become the input data for the first layer of processing nodes. The initial input values on the incoming edges of the processing nodes are the numbers from the array.

The system then catalogs all potential first-level features in the input (as motifs in the quantitative arrays). The system identifies which features are recurrent, the "jaw line" feature as compared with the "square of blue cloth" feature, for example. The two features are represented as different motifs in the underlying array, a multi-cell curve pattern for the "jaw line" feature and a certain cluster of brightness values for the "square of blue cloth" feature. The system rolls-up relevant first-level features into a higher level of abstraction in the second-level (starting to assemble elements that may eventually belong to a face, a jaw line, a nose, etc.), and discards (by down-weighting) other features that are not relevant.

The digitization process works the same way for other forms of input data, similarly converting the input into a pixelized grid of data values that are read into an array, whether the input data is speech, text, audio clips, words, or images.

9.3.3 *Optimization: Backpropagation and gradient descent*

The third aspect of the operation of deep learning networks is optimization, using loss functions to identify the minimum or maximum points of a function. Loss functions are easily applied in deep learning since the problem is formulated as an s-curve shaped logistic regression. This is good since the problem of combinatorial complexity quickly arises in the trial-and-error testing of different feature sets. Although there are benefits to having a generic and modular system, it is inefficient to blindly test *all* possible permutations, slightly up-weighting and down-weighting each. Therefore, optimization techniques are used to obtain more expedient results, and constitute the largest area of research focus and practitioner experimentation.

9.3.3.1 *Backpropagation of errors*

An important advance is the proposal of backpropagation as a solution to combinatorial complexity (Rumelhart *et al.*, 1986). Backpropagation is an optimization method in which the error contribution of each node is calculated after a batch of data is processed. Both the total error and the contribution to the error at each step going backwards are computed. The technique is referred to as the backpropagation of errors method since intermediate errors are transmitted backwards through the network in subsequent cycles as a means of signaling which solutions have a higher power of potential accuracy. A variety of statistical error calculation methods are employed. Some of the most frequently used are mean squared error (MSE), sum of squared errors of prediction (SSE), cross-entropy (softmax), and softplus (a smoothing function). The error calculation method may depend on the type of network and the problem. CNNs (using classification for image recognition) are more likely to use cross-entropy for error calculation, whereas RNNs (using regression for speech and text recognition) more often use MSE.

9.3.3.2 *Gradient descent*

Related to backpropagation is another important optimization method, gradient descent. The gradient is the slope of a function, and gradient descent is a loss function for quickly finding the minimum of a function (for example, the fastest path for rain water to reach the bottom of a canyon). In gradient descent, two values are calculated for each node, the error contribution and the gradient (slope). These values are used to identify the least contributive nodes while also reducing combinatorial complexity by quickly directing the function to an optimal solution. Although the concept of loss functions is straightforward, implementation is more complicated. Networks might deploy a multi-phased approach to loss functions, just as they do for learning (beginning with unsupervised learning to characterize a data set and then transitioning to supervised learning for more rigorous analysis and prediction). Likewise, with loss functions, the first optimization phase might be using coarse-grained models with steep gradient descent to quickly identify system minima and estimate the nodes contributing the biggest errors. Then a second optimization phase might apply fine-grained methods for greater specificity, for example, with a slowly decreasing random function to descend the landscape more steadily to obtain the most efficient solution.

Techniques such as dropout and regularization are used to improve gradient descent and address known problems in statistical learning such as overfitting. In dropout, units are randomly dropped from the network during training to create thinner networks, which are then combined to approximate an unthinned network during the test phase (Srivastava *et al.*, 2014). In regularization, additional parameters are introduced into the cost function that manages gradient descent to produce a smoother result.

9.4 Challenges and Advances

9.4.1 *Generalized learning*

So far, the success of deep learning has been limited to a specific range of applications, namely supervised learning for image and text recognition. Data availability and low-cost computing resources have facilitated this success, but there is a critique that deep learning may have reached a plateau and cannot progress farther without a different paradigm. For example, Google DeepMind researchers suggest the need to extend the field with structures that allow for reasoning about the relationships between elements in a network (in more of a traditional network science approach) to learn how the nodes in a network are connected (Battaglia *et al.*, 2018).

The problem is that deep learning systems have not produced learning techniques that are generalizable beyond their initial scope. Although a computer can beat a human opponent in chess, Go, checkers, or backgammon, in each case it is a specialized machine. The AlphaGo system (after beating human players at Go in 2016) was directed at the task of Atari game play (47 classic arcade games), with the result that while the system can learn and beat human competitors at any particular game, the architecture is not portable (Karpathy, 2017). The system can only play one game at a time, and must learn the rules from scratch for each game. There is still a vast gap between human and machine intelligence in the area of generalized learning.

One reason that it is difficult to develop a theory of generalized machine learning is because deep learning is a black box. The specific details of how the technology operates in the hidden layers is unknown. A variety of theoretical explanations have been proposed from different fields. In information theory, one idea is based on information bottlenecks. Research suggests that relevant information signals may prune networks as they operate, having the overall effect of squeezing information through a bottleneck (Shwartz-Ziv & Tishby, 2017). In the area of information geometry, another theory claims that deep learning systems do not really find solutions, but merely rewrite the problem space onto the solution space. Perhaps deep learning systems do nothing more than execute a chain of simple geometric mappings from one vector space to another (Chollet, 2017). The input data vector space is re-mapped to the vector space of output solutions, but nothing new is obtained, the problem is merely represented differently. If true, this could help explain why solutions have not been generalizable.

9.4.2 *Spin glass: Dark knowledge and adversarial networks*

One technophysics application in deep learning is recent advances using spin glass models as an optimization technique in the areas of dark knowledge and adversarial networks. Spin glasses are disordered magnets with charges locked partially in one direction and partially in the other. Spin glasses are conceptually similar to semiconductors, as devices whose partial charge capacity can be selectively engaged to manipulate energy flow. The partial charge capacity can be used to model a problem with a variable energy landscape that can be minimized (directed into a funnel) for an optimal solution. This has been demonstrated in protein-folding (Bryngelson *et al.*, 1995) and also in deep learning. Loss optimization is structured as an energy function (with a Hamiltonian (Auffinger *et al.*, 2012)) and employed to determine the minimal energy landscape in deep learning networks (Choromanska *et al.*, 2015).

Dark knowledge refers to the deep learning technique of compressing the average predictions of several model runs into one result (Hinton *et al.*, 2014). The dark (unseen) knowledge is compressed into a single model which is easier to deploy. Hence, a more complex model can be used to train a simpler, compressed model. In spin glass information-theoretic terms, the dark knowledge model continues to have the same amount of entropy as layers of the network are added and compressed, but the loss function keeps improving, decreasing the error rate and obtaining better results. The same structural point is true in adversarial networks. Adversarial networks are a deep learning self-play method in which there two networks. The adversary network generates false data and the discriminator network learns whether the data are false, not by changing the structure of the network, but by manipulating the loss function (Creswell *et al.*, 2018).

9.4.3 *Software: Nonlinear dimensionality reduction*

Another technophysics application in deep learning is related to nonlinear dimensionality reduction. Datasets often have high-dimensionality, meaning separate attributes that cannot be mapped onto each other. Technically, high dimensionality is a specialized situation that arises in data analysis in which $p > n$; there are n samples and p features (Buhlmann & van de Geer, 2011). For example, in healthcare data, there could be numerous distinct variables such as blood pressure, weight, and cholesterol level that could be high-dimensional. Dimensions might include different kinds of data as well as different kinds of interpretations of the data, for example, groups, clustering, networks, interactions, and temporal relations. Complex data sets often have high-dimensionality, including biological data (gene expression microarrays, fMRI data, and clinical trial data), computational data (collaborative filtering, recommendation engine, and sentiment analysis), and climate, energy modeling, and financial market data.

The sigmoidal logistic regression is used to map complex data with various elements into one function, and a variety of dimensionality reduction methods may be applied to simplify the data as they are analyzed by the deep learning system. One is manifold learning, configuring data on a curved manifold so that they can be visualized in a lower-dimensional space (a nonlinear dimensionality reduction technique) (Pratiher & Chattoraj, 2018). Data may also be simplified through autoencoding. Autoencoding (automatic encoding) is an unsupervised learning technique (using Markov chains and principal component analysis) in which the system learns a representation or encoding scheme for data with reduced dimensionality.

9.4.4 *Software: Loss optimization and activation functions*

Loss optimization in gradient descent continues to be a focal point in contemporary research. The challenge is finding accurate minima since systems can get stuck in saddle point traps (unfavorable local minima). More sophisticated loss optimization algorithms are proposed as a solution. One such method is Hessian Free optimization (which approximates the error function by a second-order quadratic at each step and uses a conjugate gradient) (Haider & Woodland, 2018). A technophysics method for improved gradient descent is simulating annealing (Pan & Jiang, 2015). In deep learning, simulated annealing is a probabilistic technique for approximating the global optimum of a given function. Annealing refers to an analogy with thermodynamics, in the sense of cooling metals slowly for optimal strength.

In addition to loss optimization, activating the system algorithms in the first place is another important function. The standard activation is per the logistic regression sigmoid (supporting values 0–1) or related Tanh functions (supporting values between (–1) and 1). However, some systems use a ReLU (rectified linear unit) function to activate the processing neurons, which was started by the image recognition network, AlexNet. A ReLU function is not a smooth s-curve, but a rectified angle designed to accelerate network activation and speed.

9.4.5 *Hardware: Network structure and autonomous networks*

Software is one research topic and hardware and network structure are another. Research questions concern the structure of layers (for example, the advantages and drawbacks of a 1×9 versus a 3×3 layer architecture) and addressing known issues (such as early stopping and slowdowns between layers 1 and 2 (Nielsen, 2017)). One method of analyzing the interactions between layers is Deep Belief Networks. Deep belief networks use renormalization to zoom out and coarse-grain over details to identify a system state, in this case to understand the connections between the layers as opposed to between the units within each layer. This metastructure allows pre-training processing runs to assign initial probability weights to processing nodes (Lee *et al.*, 2009).

Since many optimization algorithms are still hand-coded, having the system evolve and learn the best optimization algorithms and architecture is another contemporary research topic. Neuro-evolution is the idea of the network training itself to evolve the architecture and parameters over time (Andrychowicz *et al.*, 2016). The deep learning network can be seen as a two-tiered learning system that at one level, learns from the input data (content), and at another level, learns the best parameters and architecture for its own operation and fine-tuning (form) (Miikkulainen *et al.*, 2017). The idea is to have the system itself figure out the best learning model given the content and its own capabilities (in a sort of Howard Gardner-type delineation of learning styles for deep learning) (Gardner, 1993). The network might train itself to write software, too. In the current state of machine learning, the network might be able to generate short programs (but not long ones) (Norvig, 2014). Neural networks might create new programs by combining existing modules (Reed & de Freitas, 2016), for example, learning from execution traces (logs of computer program activations) (Zhang & Gupta, 2005).

9.5 Deep Learning Applications

9.5.1 *Object recognition (IDtech) (Deep learning 1.0)*

Deep learning systems have demonstrated results in a variety of areas such as computer vision, voice recognition, speech synthesis, machine translation, game playing, drug discovery, and robotics. A new kind of functionality is furnished, IDtech, identification technology, by analogy to FinTech (financial technology) and RegTech (regulatory compliance technology). IDtech can be used to recognize and identify both physical and digital objects, and could become a standard technology feature.

9.5.2 *Pattern recognition (Deep learning 2.0)*

Beyond object recognition (IDtech), pattern recognition functionality is becoming more prominent in a next wave of deep learning applications. A key area for system-generated pattern recognition is autonomous driving. Deep driving encapsulates the idea of having a dynamic predictive processing capacity for large volumes of real-time data in vehicle modules (Chen *et al.*, 2015). Such functionality is crucial for the operation of autonomous vehicles. Another important pattern recognition application of deep learning is text mining. There are deep learning results in problems related to document classification and clustering, document summarization, web mining, and sentiment analysis (Karakus *et al.*, 2018). There are many other venues for deep learning pattern recognition applications, for example, in fraud detection, credit scoring, data center management, and diagnostic radiology.

9.5.2.1 *Health and medicine*

Health and medicine could benefit significantly from deep learning applications, in both object recognition and pattern recognition. Image recognition in radiology and medical diagnostics is the most straightforward use case, and has been proven in detecting breast tissue lesions in images and pulmonary nodules in CAT scans (Cheng *et al.*, 2016). Likewise, dermatologist-level classification of skin cancer has been shown (Esteva *et al.*, 2017). Image-reading by machines is facilitated by the fact that there may be a sharp light–dark contrast between cancerous cells and healthy cells, which can be easily identified with the pixel brightness analysis method.

Deep learning is used in microscopy imaging as well, helping to overcome challenges such as light-scattering in the 3D visualization of cell structure (Waller & Tian, 2015). Biological neural networks, which inspired artificial neural networks, are in turn themselves the target for deep learning techniques. The visual cortex is hierarchically organized with intermediate layers, and deep learning models are applied to the study of this domain (Yamins & DiCarlo, 2016). Remote medical diagnosis via image recognition is another potential deep learning health application, as a first line of care, especially for the 400 million people worldwide who the World Health Organization estimates do not have access to essential health services (WHO, 2015).

9.5.3 *Forecasting, prediction, simulation (Deep learning 3.0)*

An emerging class of more sophisticated deep learning applications focuses on forecasting prediction, and simulation. One such application is time series forecasting using RNNs. RNNs are used to analyze sequential data, with success in language modeling, handwriting recognition, and acoustic sequence recognition (Azzouni & Pujolle, 2017). In a basic example, given a phrase beginning with the words "the grocery…," a sufficiently trained RNN is more likely to predict "store" than "school" as the next word. RNNs can analyze time series data such as stock prices and provide forecasts. In autonomous driving systems, RNNs might be used to anticipate vehicle trajectories and avoid accidents.

A key feature in the sequential processing of RNNs is memory, in particular, LSTM. LSTM is the capability of the deep learning network to have a memory function such that it can identify and recall relevant sequences. The network must figure out what to forget, and what to store for short-term and long-term processing phases. LSTMs were developed to address the technical problem of vanishing gradients (Hochreiter & Schmidhuber, 1997). Vanishing gradients is the issue where backpropagated gradients (solution slopes) can quickly grow or shrink with each time step, essentially scaling exponentially over many iterations such that they explode or vanish (LeCun *et al.*, 2015). A function such as LSTM remedies the vanishing gradients problem by serving as a placeholder that remembers important values across time step iterations. LSTMs are essentially a time complexity technique, serving as a means of exploiting sequentiality.

Fluid dynamics is a technophysics inspiration for LSTM RNNs (Kolmogorov, 1941). In situations of turbulence, recursive fractal processes may be observed, such as river eddies that create new eddies half their size, that then create more new eddies at half their size and so on, in an ongoing fractal cascade. The same principles are used as one model for constructing LSTMs, in that each subsequent layer remembers data for twice as long, in a fractal-type model. A related technique in RNN LSTMs is Neural Turing Machines, a configuration that attempts to mimic the short-term memory of the human brain by linking neural networks to external memory modules (Graves *et al.*, 2014).

LSTM RNNs operate by processing an input sequence one element at a time, and maintaining a state vector in their hidden units that contains information about the history of the past elements in the sequence. This means that nonlinear functions can be approximated in forecasting (Gamboa, 2017). LSTM RNNs can learn not only temporal dependence between data elements, but also changes in temporal dependence as input sequences change over time (Brownlee, 2017). Hence, forecasting seasonality and other pattern variations might be possible. LSTM RNNs may be better than other forecasting methods at calculating smooth trends from incomplete, noisy, and discontinuous underlying data (Gers *et al.*, 2001).

LSTM RNN forecasting results have been obtained in network traffic analysis and real-time oil pipeline monitoring. Anomaly detection is an important related application that has been demonstrated in network intrusion detection (Bontemps *et al.*, 2016), and with various data sets including ECG, electrical power demand, engine data, and space shuttle valves (Malhotra *et al.*, 2015). In forecasting competitions such as the KDD Cup and the Web Traffic Time Series Forecasting competition (run on competition platforms Kaggle, CrowdAnalytix, and Tunedit), hybrid methods employing both traditional statistical models and machine learning methods are often used.

9.5.3.1 *CNNs for time series forecasting*

Although the sequential estimation techniques used by RNNs correspond most obviously to forecasting, CNNs might also make a contribution. CNNs are used for image recognition, in which an important feature that might be distinguished is symmetry, for example, in facial recognition (Saber & Tekalp, 1998). In genomic data, there are symmetries between alternating areas of exons (coding regions) and introns (non-coding regions), and thus one application of CNNs could be identifying symmetry pattern disruptions in genomic analysis that might indicate disease-causing mutations (Leung *et al.*, 2015). The known temporal dimension of how such mutations develop could be incorporated in predictive and preventive medical care strategy.

Symmetry is an important property in many domains of sequential data and is also a property of complex systems. Symmetry-breaking can be important in signaling phase transitions within a system (Girault, 2015). Hence, deep learning CNNs might be applied in a time series forecasting model to recognize symmetry and symmetry-breaking within systems over time.

References

Andrychowicz, M., Denil, M., Gomez. S. *et al.* (2016). Learning to learn by gradient descent by gradient descent. arXiv:1606.04474 [cs.NE].

Auffinger, A., Ben Arous, G. & Cerny, J. (2012). Random matrices and complexity of spin glasses. *Comm. Pure Appl. Math.* 66(165):165–201.

Azzouni, A. & Pujolle, G. (2017). A long short-term memory recurrent neural network framework for network traffic matrix prediction. arXiv:1705.05690 [cs.NI].

Battaglia, P.W., Hamrick, J.B., Bapst, V. *et al.* (2018). Relational inductive biases, deep learning, and graph networks. arXiv:1806.01261 [cs.LG].

Bontemps, L., Cao, V.L., McDermott, J. & Le-Khac, N.-A. (2016). Collective anomaly detection based on long short-term memory recurrent neural networks. In T.K. Dang *et al.* (eds). *FDSE 2016, LNCS 10018*. Cham, Switzerland: Springer International Publishing, pp. 141–52.

Brownlee, J. (2017). The promise of recurrent neural networks for time series forecasting, long short-term memory networks. In *Master Machine Learning Algorithms*. Available online at http://machinelearningmastery.com/promise-recurrent-neural-networks-time-series-forecasting/.

Bryngelson, J.D., Onuchic, J.N., Socci, N.D. & Wolynes, P.G. (1995). Funnels, pathways, and the energy landscape of protein folding: a synthesis. *Proteins: Struct., Funct., Bioinf.* 21(3):167–95.

Buhlmann, P. & van de Geer, S. (2011). *Statistics for High-dimensional Data: Methods, Theory and Applications*. Heidelberg, Germany: Springer.

Burn-Murdoch, J. (2012). Study: Less than 1% of the world's data is analysed, over 80% is unprotected. *The Guardian*.

Carrasquilla, J. & Melko, R.G. (2016). Machine learning phases of matter. arXiv:1605.01735 [cond-mat.str-el].

Chen, C., Seff, A., Kornhauser, A. & Xiao, J. (2015). DeepDriving: Learning affordance for direct perception in autonomous driving. In *ICCV Proceedings*, pp. 2722–30.

Cheng, J.Z., Ni, D., Chou, Y.H. *et al.* (2016). Computer-aided diagnosis with deep learning architecture: Applications to breast lesions in US images and pulmonary nodules in CT scans. *Nat. Sci. Rep.* 6(24454).

Chollet, F. (2017). *Deep Learning with Python.* Grand Forks, ND: Manning Publications.

Choromanska, A., Henaff, M., Mathieu, M. *et al.* (2015). The loss surfaces of multilayer networks. *Artificial Intelligence and Statistics* 38:192–204.

Cleveland, W.S. (2001). Data science: An action plan for expanding the technical areas of the field of statistics. *Intl. Stat. Rev.* 69(1):21–6.

Creswell, A., White, T., Dumoulin, V. *et al.* (2018). Generative adversarial networks: An overview. *IEEE Signal Process.* 35(1):53–65.

Esteva, A., Kuprel, B., Novoa, R.A. *et al.* (2017). Dermatologist-level classification of skin cancer with deep neural networks. *Nature* 542:115–8.

Gamboa, J. (2017). Deep learning for time-series analysis. University of Kaiserslautern, Kaiserslautern, Germany. Seminar on Collaborative Intelligence in the TU Kaiserslautern. arXiv:1701.01887 [cs.LG].

Gao, H. (2017). A Walk-through of AlexNet. Medium.

Gardner, H. (1993). *Multiple Intelligences: The Theory in Practice.* New York, NY: Basic Books.

Gers, F.A., Eck, D. & Schmidhuber, J. (2001). Applying LSTM to time series predictable through time-window approaches. In *International Conference on Artificial Neural Networks. ICANN 2001*: *Artificial Neural Networks*, pp. 669–76.

Girault, J.M. (2015). Recurrence and symmetry of time series: Application to transition detection. *Chaos. Soliton. Fract.* 77:11–28.

Graves, A., Wayne, G. & Danihelka, I. (2014). Neural turing machines. arXiv 1410.5401 [cs.NE].

Haider, A. & Woodland, P.C. (2018). Combining natural gradient with Hessian free methods for sequence training. arXiv:1810.01873 [cs.LG].

Halevy, A., Norvig, P. & Pereira, F. (2009). The unreasonable effectiveness of data. *IEEE Intell. Syst.* 24(2):8–12.

He, K., Zhang, X., Ren, S. & Sun, J. (2015). Delving deep into rectifiers: Surpassing human-level performance on ImageNet classification. In *ICCV '15 Proceedings of the 2015 IEEE International Conference on Computer Vision (ICCV)*. December 7–13, pp. 1026–34.

Hinton, G., Vinyals, O. & Dean, J. (2014). Distilling the knowledge in a neural network. NIPS 2014 Deep Learning workshop. arXiv:1503.02531 [stat.ML].

Hinton, G.E. (2007). Learning multiple layers of representation. *Trends Cogn. Sci.* 11(10):428–34.

Hochreiter, S. & Schmidhuber, J. (1997). Long short-term memory. *Neural Comput.* 9(8):1735–80.

Huang, G., Sun, Y. & Liu, Z. (2016). Deep networks with stochastic depth. arXiv:1603.09382v1. https://arxiv.org/abs/1603.09382v1.

Karakus, B.A., Talo, M., Hallac, I.R. & Aydin, G. (2018). Evaluating deep learning models for sentiment classification. *Concurr. Comp*. Wiley. 30(21):e4783.

Karpathy, A. (2017). AlphaGo, in context. Medium.

Kolmogorov, A.N. (1941). The local structure of turbulence in incompressible viscous fluids at very large Reynolds numbers. *Dokl. Akad. Nauk. SSSR.* 30:301–5 [Reprinted 1991: *Proceedings of the Royal Society of London. Series A*. 434:9–13].

Krizhevsky, A., Sutskever, I. & Hinton, G.E. (2012). ImageNet classification with deep convolutional neural networks. Proceedings NIPS 2012. In *Proceedings of the 25th International Conference on Neural Information Processing Systems*, Volume 1. Lake Tahoe, Nevada. December 3–6, pp. 1097–105.

Le, Q.V., Ranzato, M., Monga, R. *et al.* (2012). Building high-level features using large scale unsupervised learning. arXiv:1112.6209 [cs.LG].

LeCun, Y., Bengio, Y. & Hinton, G. (2015). Deep learning. Nature. 521:436–44.

Lee, H., Grosse, R., Ranganath, R. & Ng, A.Y. (2009). Convolutional deep belief networks for scalable unsupervised learning of hierarchical representations. In *ICML '09 Proceedings of the 26th Annual International Conference on Machine Learning*. Montreal, Quebec, Canada. June 14–18, pp. 609–16.

Leung, M.K.K., Delong, A., Alipanahi, B. & Frey, B.J. (2015). Machine learning in genomic medicine: A review of computational problems and data sets. *Proc. IEEE 2015.* 104(1):176–97.

Malhotra, M., Vig, L., Shroff, G. & Agarwal, P. (2015). Long short term memory networks for anomaly detection in time series. In *ESANN 2015 Proceedings, European Symposium on Artificial Neural Networks, Computational Intelligence and Machine Learning*. Bruges, Belgium. April 22–24.

Miikkulainen, R., Liang, J., Meyerson, E. *et al.* (2017). Evolving deep neural networks. arXiv: 1703.00548 [cs.NE].

Nielsen, M. (2017). Improving the way neural networks learn. In *Neural Networks and Deep Learning*. San Francisco CA: Determination Press.

Norvig, P. (2014). Machine learning for programming. In *SPLASH '14 Proceedings of the Companion Publication of the 2014 ACM SIGPLAN Conference on Systems, Programming, and Applications: Software for Humanity*. Portland, Oregon. October 20–24, pp. 1–73.

Oxford Economics (2018). Workforce 2020. https://www.oxfordeconomics.com/workforce2020. Accessed June 30, 2019.

Pan, H. & Jiang, H. (2015). Annealed gradient descent for deep learning. Proceedings: 31st Conference on Uncertainty in Artificial Intelligence (UAI), Amsterdam, Netherlands. July 12–16.

Pratiher, S. & Chattoraj, S. (2018). Manifold learning & stacked sparse autoencoder for robust breast cancer classification from histopathological images. arXiv:1806.06876v1 [cs.CV].

Reed, S. & de Freitas, N. (2016). Neural programmer-interpreters. arXiv: 1511.06279v4 [cs.LG].

Rumelhart, D.E., Hinton, G.E. & Williams, R.J. (1986). Learning representations by back-propagating errors. *Nature* 323:533–6.

Saber, E. & Tekalp, A.M. (1998). Frontal-view face detection and facial feature extraction using color, shape and symmetry based cost functions. *Pattern Recognit. Lett.* 19(8):669–80.

Sherbin, B. (2017). Jensen Huang Keynotes NVIDIA's 2017 GPU Technology Conference. NVIDIA blog.

Shwartz-Ziv, R. & Tishby, N. (2017). Opening the black box of deep neural networks via information. arXiv:1703.00810 [cs.LG].

Srivastava, N., Hinton, G., Krizhevsky, A. *et al.* (2014). Dropout: A simple way to prevent neural networks from overfitting. *J. Mach. Learn. Res.* 15: 1929–58.

Swan, M. (2017). Is Technological unemployment real? abundance economics. In: Hughes, J., LaGrandeur, K. (eds). *Surviving the Machine Age: Intelligent Technology and the Transformation of Human Work*. London: Palgrave Macmillan, pp.19–33.

Trask, A. (2018). *Grokking Deep Learning*. New York, NY: Manning Publications.

Varian, H.R. (2014). Big data: New tricks for econometrics. UC Berkeley. Available online at: http://people.ischool.berkeley.edu/~hal/Papers/2013/ml.pdf.

Waller, L. & Tian, L. (2015). Computational imaging: Machine learning for 3D microscopy. *Nature* 523:416–7.

Wang, L. (2016). Discovering Phase Transitions with Unsupervised Learning. *Phys. Rev. B*. 94(19):195105.

WHO (World Health Organization). (2015). New report shows that 400 million do not have access to essential health services. WHO and World Bank. http://www.who.int/mediacentre/news/releases/2015/uhc-report/en/.

Yamins, D.L.K. & DiCarlo, J.J. (2016). Using goal-driven deep learning models to understand sensory cortex. *Nat. Neurosci.* 19(3):356–65.

Zhang, X. & Gupta, R. (2005). Whole execution traces and their applications. *ACM T Archit. Code Op.* 2(3):301–34.

Chapter 10

Quantum Machine Learning

Abstract

One of the first uses of quantum computing is machine learning. Both techniques are aimed at the same kinds of problems in statistical data analysis and optimization. In this sense, quantum computing may be an improved method over machine learning. Machine learning applications such as optimization and simulation are being instantiated in quantum computing formats. Standardized tools are emerging such as the variational quantum eigensolver (VQE) and the quantum approximate optimization algorithm (QAOA). Quantum algorithms are being specified with standard gate logic, including the Hadamard gate (which acts on 1 qubit to put it in a superposition state), the CNOT gate (which acts on 2 qubits to flip one), and the Toffoli gate (which acts on three or more qubits to implement the Boolean operators). Other advances in geometric deep learning and information geometry facilitate a wider range of 3D computational operations in problem-solving.

10.1 Machine Learning, Information Geometry, and Geometric Deep Learning

10.1.1 *Machine learning as an n-dimensional computation graph*

Machine learning is already quantum-ready (ready for implementation in quantum information systems) in several ways. First, machine learning operates on computation graphs, which are a 3D format (any graph implies 3D). Second, relevant operations in machine learning are migrating to tensor networks, the analytic tool of choice for modeling quantum information problems. Third, there are novel formulations such as information geometry and geometric deep learning (deployed in graph convolutional networks, GCNs). These advances allow a wider range of non-Euclidean geometries to be engaged in deep learning problems (geometry is a selectable parameter), and information is analyzed in its native form rather than being compressed into lower-dimensional representations for processing by the deep learning system.

10.1.1.1 *Tensor networks for machine learning*

Machine learning methods are based on linear algebra applied in matrix multiplications. The matrix multiplications multiply various ranges of scalars, vectors, matrices, and tensors. A scalar is a single number. A vector is representable by an array of multiple numbers arranged in order, with each number identified by its index in that ordering. A matrix is representable as a 2D array of numbers, with each element identified by two indices. A tensor is representable by an array of numbers arranged on a regular grid with a variable number of axes identified by indices. Efficient hardware for processing matrix multiplications extends from CPUs to GPUs and tensor-processing units (TPUs). TPUs, are operated in clusters with TensorFlow software (an open source machine learning library). The specialized hardware and software allow the machine learning algorithms to flow through the vast series of matrix multiplications without having to store intermediate values in memory.

Tensor networks refer to a specific technophysics method that is starting to be implemented to a greater degree in machine learning, particularly in the contemplation of quantum computing (Huggins *et al.*, 2019). A tensor network is a collection of tensors with indices connected according to a network pattern (as in the usual method of tensor-based machine learning), but in particular, it is used to efficiently represent quantum mechanical problems such as a many-body wave-function in an exponentially large Hilbert space (functionality that has not always been needed for machine learning problems).

10.1.1.2 *Dimensionality reduction*

A problem in machine learning and beyond is dimensionality reduction. The concept of dimensionality is widely used in data science and machine learning to refer to the overall size and complexity of a dataset, but dimensionality also has a specific technical definition. In the analytically precise definition, the number of dimensions correlates with the number of features (attributes) per example (datapoint) in a dataset. The term *high dimensionality* is a specialized situation that arises in data analysis in which $p > n$; there are n samples and p features (Buhlmann & van de Geer, 2011). The dataset has more features than samples.

This could be true for datasets of any size. For example, a dataset with 3 data points and 5 features constitutes high-dimensional data, whereas another dataset with 500,000 features and 1 million samples is low-dimensional data, based on the technical definition. A canonical example of high-dimensional data occurs in genomic analysis, for example, having a dataset of $n = 1500$ people and $p = 1$ million SNP (polymorphism) locations in the genome. Healthcare data are known for having large numbers of variables (e.g. height, weight, blood pressure, cholesterol level), but unless $p > n$, it is not high-dimensional data. Datasets with high numbers of variables may still be difficult to process, but they are not as computationally difficult as high-dimensional data. High-dimensional data are difficult to treat statistically because there cannot be a deterministic answer when $p > n$ unless additional assumptions are introduced into the model.

A key complexity technology (ComplexityTech: A tool for managing complexity) is dimensionality reduction. Dimensionality reduction is a challenge in processing datasets, whether they are large and complicated or specifically high-dimensional. The idea of dimensionality reduction refers to any variety of techniques that are used to simplify the structure of complex datasets while maintaining data integrity. Initially, only linear models such as principal component analysis and classical metric multidimensional scaling were available. However, since the 1990s, nonlinear dimensionality reduction methods have been a substantial advance (Lee & Verleysen, 2007). Nonlinear approaches may be applied to both problem representation and analysis.

One nonlinear dimensionality reduction method is manifold learning, in which data are configured on a curved manifold so that they can be visualized in a lower-dimensional space. Such visualization techniques use graphs to represent the manifold topology of data problems, which introduces new computational metrics such as geodesic distance. Other methods are being used for improved optimization in the context of dimensionality reduction, for example, algorithms based on kernel techniques and spectral decomposition. This has led to a new method, spectral embedding, which may outperform previous tools. Even with the new nonlinear dimensionality reduction techniques, though, certain problems in machine learning are not yet resolved such as the so-called "curse of dimensionality". The curse of dimensionality refers to the situation in which the performance of machine learning algorithms spikes at first, but then levels off after a certain number of features are added. The reason is not understood, but might be explored in more sophisticated quantum machine learning models.

10.1.2 *Information geometry: Geometry as a selectable parameter*

One geometry cannot be more true than another; it can only be more convenient

Henri Poincaré (1905, p. 59)

Just as there are different kinds of space in mathematics and physics, there are different kinds of geometry. As mathematician Poincaré indicates (when talking about non-Euclidean geometries), there is no one true geometry, only different geometries that may be more or less convenient for the problem at hand. The main kinds of geometries are Euclidean geometry (familiar everyday geometry) and non-Euclidean geometry such as spherical and hyperbolic geometry. Spherical geometry is geometry on the edge of a sphere, and hyperbolic geometry is the geometry on saddle surfaces. Riemannian geometry (also called elliptical geometry), is another form of non-Euclidean geometry, based on differential calculus that is relevant for calculating spacetime curvature.

The framing concept in information geometry is selective geometries, the idea that geometry can be chosen as a parameter as any other, perhaps in the same way that the space and time regimes can be selectable parameters of a physics problem. In machine learning, there could be a choice of geometries which pertain to the kind of geometrical space in which the optimization is performed. Different geometries might be better for different applications. One parameter-related choice could be simplification, for example, discretizing high-dimensional continuous geometries to simplified discretized geometries for useful computation.

Information geometry is an interdisciplinary field that considers geometry in the context of information science (Amari, 2016). Differential geometry techniques are applied to the study of probability theory and statistics. One focus is statistical manifolds, which are Riemannian manifolds whose points correspond to probability distributions. Geometry (whether assumed or selected) is a foundation for probabilistic modeling, in that the underlying geometry can have a substantial influence on how the optimization or statistical modeling problem is solved. Information geometry enables new kinds of more sophisticated machine learning techniques that take advantage of the properties of non-Euclidean geometry. For example, a deep learning problem could be instantiated as an optimization problem defined over a statistical manifold in which points are probability distributions. One effort in information geometry and deep learning is the DeepRiemann project (Malago *et al.*, 2018). Riemannian optimization methods and differential geometry are applied towards the development of improved deep learning algorithms.

10.1.3 *Geometric deep learning*

Geometric deep learning is a deep learning method in which algorithms analyze information in its native form, rather than translating it into a lower-dimensional representation to be processed by the system. The insight is that input data (images, text, audio, or otherwise) often exist naturally in higher-dimensional forms. Data are compressed to be conducive to processing by machine learning algorithms, which may destroy the very relationships between data elements that the system is attempting to detect. Hence, the extent to which the technology, deep learning networks, can process data natively may improve results.

One means of implementing geometric deep learning is through GCNs. In one example, semi-supervised learning is conducted on graph-based data using a version of convolutional neural networks that operate directly on graphs (Kipf & Welling, 2017). The method is congruent with the data. The model scales linearly in the number of graph edges and learns hidden layer representations that encode both local graph structure and the features of the nodes. The results outperform traditional methods.

Another way that geometric deep learning is employed is in trying to understand why deep learning has been so successful at certain machine learning tasks such as image recognition. Deep learning is a proven method, but is also a black box technology, in that specifically how it works is not understood. Work involving geometric deep learning purports that the success may be due to a manifold structure existing in the underlying data (Lei *et al.*, 2018). The claim is that in its native representation, high-dimensional data may be naturally concentrated close to a low-dimensional manifold. A geometric deep learning algorithm can be used to learn the shape of the manifold and the probability distributions on it. The point is that improved computational methods that are more in line with the natural attributes of phenomena may be better at detecting the patterns and structure present in the phenomena. Other proposals might be similarly tested with a geometric view, such as the idea that the success of deep learning is due to a pruning mechanism in the deep learning network that forces information through a bottleneck (Shwartz-Ziv & Tishby, 2017).

10.2 Standardized Methods for Quantum Computing

Machine learning problems are among the first applications of quantum computing that are currently being demonstrated. A standardized set of computational methods (tools and algorithms) is emerging, intended for use on any quantum computing platform. The first kinds of tasks that are being investigated involve the quantum instantiation of machine learning techniques directed toward optimization and simulation problems.

10.2.1 *Standardized quantum computation tools*

A standardized set of tools and algorithms is starting to be established for quantum computing. The Bernstein–Vazirani algorithm is the "Hello, World!" equivalent in quantum computing, in that it is one of the first basic demonstrations performed in whatever form of quantum computing hardware is being used. The Bernstein–Vazirani algorithm is used to extract certain specific encoded bits from a string (Bernstein & Vazirani, 1997). The team contributed important early work regarding the computational complexity of quantum computers and classical computers, quantifying the potential benefits of quantum computing.

A second standardized tool that is emerging in quantum computation is the variational quantum eigensolver (VQE) (Peruzzo *et al.*, 2014). The VQE is a quantum-classical hybrid algorithm that can be used to find the eigenvalues of a matrix. The method was initially developed in the context of quantum chemistry. An eigensolver is a program designed to calculate solutions to 3D problems (for example, in quantum mechanical systems or in big data problems with multiple dimensions). In the quantum computing context, the VQE is well suited to solving certain classes of problems on near-term NISQ devices. The strategy of the method is that even minimal quantum resources are useful when combined with classical routines. VQEs have been demonstrated in many kinds of quantum computation problems in optimization and simulation, for example, in finding the ground state energy of a Hamiltonian in an energy landscape optimization problem.

A third important standardized tool is the quantum approximate optimization algorithm (QAOA) proposed by (Farhi *et al.*, 2014) and developed into a readily-usable format as a quantum alternating operator ansatz (guess) (Hadfield *et al.*, 2019). The QAOA is designed to work on combinatorial optimization problems. The algorithm uses a QAOA to solve an optimization routine, for example, in providing solutions to the Traveling Salesman Problem. Computationally, the QAOA is a polynomial time algorithm for finding a "good" solution to an optimization problem (meaning finding an acceptable answer in a reasonable amount of time (polynomial time)). Graph theoretically, the QAOA is a form of a Max-Cut solution on regular graphs in which the quantum algorithm finds a cut that is as close to the optimal cut as possible.

Max-Cut (maximum cut) refers to using partition functions to find the best division or optimal cut through a set of data (the best way to cut a data plot in half, for example). Instead of checking every point, the idea is to use an optimization algorithm to efficiently divide the data, finding the maximum cut that is closest to the actual optimal cut or partition through the data. Quantum computers could likely execute the calculation more quickly than classical computers.

There are different ways of setting up the Max-Cut problem in quantum computers to apply the QAOA. One approach is instantiating the Max-Cut problem as an energy optimization. The implementation constructs a physical system (in a quantum annealing machine), typically with a set of interacting spin particles, whose lowest energy state encodes the solution to the problem, so that solving the problem is equivalent to finding the ground state of the system. An QAOA is used to identify the energy landscape minimum. Another approach, in graph theory, involves starting with a set of unlabeled points and assigning labels based on a similarity metric (such as Euclidean distance). Then, the points are mapped onto a planar graph, and treated as a Max-Cut problem. A QAOA is applied to solve through an optimization routine.

10.2.2 *Standardized quantum computation algorithms*

In addition to standardized tools (the VQE and the QAOA), standardized quantum algorithms are also necessary for the successful implementation of quantum computation. The first step in the DiVincenzo Criteria of overall standards for quantum computers is having a reliable system for making qubits (DiVincenzo, 2000). The next four items involve manipulating the qubits into executable computations. The four criteria are initializing the qubits, using universal quantum gates, computing with the qubits in a coherence time that is long enough to be useful, and measuring the result at the end.

Although there are many ways of making qubits, once they are produced, there is more of a standardized approach for computing with them. Whatever hardware method is used to produce the qubits, a quantum chip is typically the result. The quantum computing chip can then be used to solve problems with a set of standard quantum algorithms. The aim is to have generalized algorithms and programming models that can be used to design programs for any quantum chip or architecture. Quantum programmers program the quantum algorithms and access quantum computers through an internet-based interface, either locally or as a cloud service. Initially, classical algorithms were being run on quantum computers to demonstrate quicker problem-solving. More recently, there is a shift to developing native quantum algorithms to take advantage of the specific properties of quantum systems.

10.2.2.1 *Quantum logic gates and circuit diagrams*

In classical computing, information units (bits) are passed through a series of gates in a logic circuit to perform a series of computational operations. Likewise, in quantum computing, gates are applied to transform information units (qubits) as they proceed sequentially through a logic circuit. In quantum logic circuit design, a circuit diagram of the quantum gate architecture may have horizontal lines representing the qubits and vertical lines representing the gates. For example, nine horizontal lines would indicate a 9-qubit system. The circuit diagram reflects the physical architecture of the chip that has nine superconducting loops, each passing current in both directions around the loop to create the qubits (creating the two energy states of the qubit with a spinning atom in each loop). In the circuit diagram, the qubit lines are represented horizontally, and gates are represented vertically, each gate crossing two or three of the qubit lines. To conduct a computation, a transformation is applied to the gates. The circuit is executed linearly in time, from left to right according to the circuit diagram.

Unlike in the classical system, it is not possible to provide independent descriptions of the states of each of the 9 qubits in the quantum information system. Instead, a vector is used to describe the state of the quantum system. The size of the vector is 2^n for an n-qubit system, so a

9-qubit system would have a vector size of 512. To begin the computation, the starting vector of 512 numbers is initialized to a particular state, typically all 0s. Then the computation runs through the gates in the circuit. Each gate is applied to the qubits and the vector is updated with the output values of each operation. At the end of the computation, a measurement is made on the final vector, and the results are read into classical output values (0s and 1s) for downstream treatment in classical applications.

10.2.2.2 *Superposition and quantum logic gates*

The superposition property of qubits operates in the context of quantum logic gates. A qubit exists as a 0 <u>and</u> a 1 until it is collapsed in measurement to a classical bit and becomes a 0 <u>or</u> a 1. The qubit exists in a 3D space called a Hilbert space. (The Hilbert space can grow to arbitrarily many dimensions as more qubits are added, but the basic Hilbert space for 1 qubit is a 3D space like the everyday physical world.) The qubit can be anywhere in the 3D space, along the *X*, *Y*, and *Z* axes, and is moving around in this space. The state of the qubit is a vector in the 3D space, represented by an arrow, indicating a location with position and momentum.

The vector has an orientation in the 3D space and this is how quantum logic gates, or quantum operations, are applied to it. Quantum gates are matrices that are multiplied by the vectors to move the vector around in the 3D space. For example, an initial quantum state vector that points up and right, after being multiplied by the first quantum gate, might point up and left. Each gate operation is a new multiplication that continues to move the quantum state vector around in the 3D space. Measurement happens at the final moment of the computation, collapsing the quantum state vector into a classical bit of 0 or 1, to whichever basis (0 or 1) the vector is in closest proximity. Vectors pointing in the upper half of the 3D space (from the "equator" to the "North pole") toward the 0 basis are collapsed into a classical information bit of 0, and those in the lower half of the 3D space are collapsed to 1. Measurement is the linear algebra sum of the series of vector manipulations through the quantum gates operations. This is how the algorithmic model of quantum computation is composed. The power of quantum computing comes from the exponential space of possible quantum states (qubits exist in a full 3D Hilbert space until collapsed and measured).

An important point is that in the standard gate model of quantum computing, many aspects are controllable, particularly the input states of the qubits (which can be initialized to a state of zero) and how they are coupled and can interact. In quantum algorithm design, the aim is to design gates in such a way that moves the vectors around to solve a problem of interest. The vectors should be aligned so that when measured, there is a high probability of obtaining a useful answer to the problem, and a reduced probability of getting a wrong or useless answer.

10.2.2.3 *Quantum computing gates*

The three basic quantum computing gates are the Hadamard gate, the CNOT gate, and the Toffoli gate. Respectively, the gates act on one, two, and three or more qubits. The Hadamard gate creates the superposition of qubits, the CNOT gate can flip the bits, and the Toffoli gate can execute all of the regular Boolean operations of a classical circuit. A qubit is initialized to a 1 or a 0 when it enters the quantum circuit (typically, all qubits are initialized to 0). Since the qubits all start in the 0 state, a transformation needs to be applied so that a qubit enters a superposition state of 0 and 1, and this is accomplished with a Hadamard gate. When a Hadamard gate is applied to a qubit, the qubit becomes a superposition of 0 and 1. The qubit is essentially now a vector in the Hilbert space. The application of the Hadamard gate means that the qubit is in a position that is the sum of the 0 vector and the 1 vector, with a 50% change of being in either, or essentially being in all possible states, in the 3D space.

The Hadamard gate acts on 1 qubit to put it in a superposition state and the CNOT gate acts on 2 qubits. The CNOT (controlled NOT) gate is a simple IF statement operating in a 2-qubit system that flips the *target* bit if the *control* bit is 1. The instruction is that in a pair of qubits, if the control bit is 0, nothing happens to the target bit. If the control bit is 1, the target bit is flipped (a 0 becomes a 1 or vice versa). The CNOT gate starts to introduce a basic computational instruction set into the quantum logic gate with IF/THEN logic.

The Hadamard and CNOT gates are specific to the quantum domain, the Toffoli gate is a classical logic gate with a direct implementation in the quantum environment. The Toffoli gate acts on three or more qubits to implement the regular Boolean operators. The Boolean operators are the six logical functions that are at the heart of classical logical operations (AND, conditional AND, OR, conditional OR, exclusive OR, and NOT).

The important indication is that quantum circuits are able to perform all of the same operations as classical circuits. However, the Toffoli gate in the quantum environment may need to rely on ancilla (ancillary) qubits to complete its operations and to perform error correction. Since it is difficult to measure and work with qubits directly in quantum systems, an ancilla is used. Ancilla qubits take advantage of the property of qubits being in entangled states with one another. Entanglement means that the data of 1 qubit can be mapped onto other qubits with which it is entangled. The data can be operated on, and checked, with the ancilla qubits, without damaging the original qubit. Beyond Hadamard, CNOT, and Toffoli gates, other gates specific to the quantum domain include Fredkin, Ising, and Deutsch gates.

10.2.3 *Quantum optimization*

Optimization means the best or most effective way to carry out a task, and is the first application in which quantum computing may offer an advantage over classical computing. In computing networks and decision science, a canonical optimization challenge is the Traveling Salesman Problem. The problem is that given a list of cities and the distances between each pair of cities, find the shortest possible route to visit each city and return to the original city. In terms of computational complexity, the Traveling Salesman Problem is an NP-hard problem.

In both classical and quantum systems, optimization is modeled in many different ways. One technique is framing optimization as a minimization problem. This could be in the algorithmic domain as an error minimization problem (for example, deep learning neural networks are measured by error rate), or in the physical domain as a minimal energy landscape problem (for example, annealing and spin glass models find the most stable energy configurations). Quantum annealing is one of the most robust models that has been used in quantum computation to solve optimization problems.

Quantum annealing-based optimization has many practical demonstrations. For example, Volkswagen has used a D-Wave quantum annealing machine to optimize travel time in Beijing for a network of 418 taxis, and intends to implement the resulting traffic management system in Lisbon (Hetzner, 2019). The same kind of large-scale quantum optimization is used in AdTech for web browser promotion placement, and in UK online grocer Ocado's automated warehouses for scheduling. The optimization algorithm coordinates hundreds of robots passing within 5 mm of each other at speeds of 4 m/s, in the process of fulfilling 65,000 orders per week (Fassler, 2018).

10.2.3.1 *Optimization: Quantum annealing*

Quantum annealers are physical devices that attempt to solve NP-complete optimization problems by exploiting quantum mechanical properties. The basic principle of quantum annealing is to encode the optimization problem in Ising interactions (ferromagnetic phase transitions) between qubits (Lechner *et al.*, 2015). Quantum annealing is emblematic of framing computational problems as an energy landscape. The way that quantum annealing (or simulated annealing more generally in classical computing) is used in optimization problems is to solve the problem of local minima in gradient descent. In gradient descent, an algorithm selects a random point in the landscape, checks the height of near-by neighbors, and descends in the steepest slope indicated by the local landscape. The problem is that although gradient descent may find local minima points, these may not be the global minimum (and therefore, the best answer for the optimization problem). Annealing is a technique that is able to scour the entire landscape by going up and down all of the peaks in the landscape to find the global minimum (as opposed to only being directed downwards as in basic gradient descent algorithms).

In quantum annealing, qubits exist in a superposition state of being 0 and 1 at the same time. This means that the quantum solver is not traversing the landscape point by point (all X and Y coordinates), but rather that the quantum solver exists in a 3D blanket-like state spread over the entire problem space. The landscape comprises all of the qubits in the system. The blanket or fog layer lying over the landscape is a direct analogy to the quantum wave function. As the annealing cycle runs, the fog layer condenses to reveal one point at the global minimum of the landscape. All of the qubits participate in the annealing cycle, with their spins flipping back and forth until each settles into one of the two possible final states (0,1), which overall indicates the lowest-energy state of the system. The quantum annealer is programmed to find gaps in the annealing trajectory between the ground state and the first excited state, and the second excited state, and so on, as the mechanism for identifying the lowest-energy configuration of the system.

10.2.3.2 *Quantum annealing Hamiltonian*

Hamiltonians are functions used to measure energy landscapes, and hence feature prominently in quantum annealing algorithms. A quantum annealing problem is an energy function optimization. In general, a Hamiltonian is a function that provides a mathematical description of a physical system in terms of energy. The input to the Hamiltonian is information about the system state, and the output is the energy of the system. There are Hamiltonians for both quantum systems and classical systems. In the quantum computing context, the input to the Hamiltonian is the state of all of the qubits in the system, and as usual, the output is the total energy of the system. The quantum annealing process is an attempt to determine the state of the qubits that produces the lowest-energy state of the system, and this is checked by inputting the proposed system configuration into the quantum Hamiltonian.

In classical computing, the Hamiltonian used in the optimization problem of finding the energy landscape minimum is a probabilistic sum of the values in the landscape. The output of the classical Hamiltonian is a specific point value. In quantum computing on the other hand, the algorithm in the optimization calculation is a quantum Hamiltonian. The quantum Hamiltonian still calculates the total energy in the system, but the quantum Hamiltonian is an operator on the multi-dimensional Hilbert space of qubits. As a quantum operator, the quantum annealing Hamiltonian has two elements, a longitudinal field and a transverse field.

The longitudinal field and the transverse field correspond to the two main kinds of waves, longitudinal (along the direction of travel) and transversal (perpendicular to the direction of travel). Quantum mechanical systems comprise vibrating particles that behave according to the Schrödinger wave function, and are modeled as such. The equation for the quantum Hamiltonian has both a longitudinal wave component and a transversal wave component. The longitudinal wave portion corresponds to the potential energy of the system. The transversal wave portion corresponds to the quantum fluctuations that are used, and then settled down in the annealing process.

Each wave component can be seen as a field. There is the longitudinal field that corresponds to the energy terms, and the transverse field that corresponds to the quantum fluctuations or annealing terms. The two terms in the quantum Hamiltonian are operators on the two different fields. In the quantum Hamiltonian, the longitudinal term is an operator acting on the longitudinal field (energy field). The transverse term is an operator acting on the transverse field (related to the quantum fluctuations in annealing). Since the two terms in the Hamiltonian are operators (matrices), solving for the Hamiltonian yields an even bigger matrix (with many hidden tensor products). Ultimately, the Hamiltonian corresponds to the optimal energy landscape of the whole system.

The point is that the quantum annealer incorporates the wave movement of quantum particles by having two different terms in the quantum Hamiltonian, the longitudinal field term (an energy term), and the transverse field term (an annealing term). Then, the problem is solved as a straightforward quantum annealing problem. The answer corresponds to the optimal (lowest-energy) landscape of the system.

Beyond the basic quantum annealing model are more advanced operations. These include reverse annealing (which entails starting from the end of the annealing cycle and going backwards into the quantum regime) and sawtooth annealing (applying a sawtooth-patterned approach to make the annealing more efficient). Besides basic optimization, quantum annealing can also be used to solve partition function problems such as the Max-Cut problem and graph coloring.

10.2.4 *Quantum simulation*

Optimization is one class of problems that may have improved problem-solving possibilities in quantum computing environments, and simulation is another. A concept in simulation is digital twin technology. The idea is that in an advanced computational society, all systems of sufficient complexity are simulated (have a digital twin). This already includes national defense systems, satellite networks, transportation systems, global financial markets, supply chains, food webs, aircraft design, and building HVAC and security systems. In the farther future, given greater computational capacity, there could be a digital twin for each individual person for data collection related to lifelogging and health management. In some ways, simulation is a more complicated version of optimization, with additional factors and scenarios, including the evolution of system dynamics over time. The main objective is to use quantum computers to simulate chemical, physical, and biological systems to understand and replicate how they operate, and to generate new materials, processes, and pharmaceutical substances as a result. Any kind of complex quantum simulation will likely require millions of qubits and error correction.

10.2.4.1 *Nitrogen fixation*

Nitrogen fixation is a high-profile problem that is targeted for potential resolution with quantum computing. Nitrogen fixation refers to understanding the natural process of biological nitrogen fixation in nitrogenase, which is fixing or assimilating nitrogen into organic compounds. This could be important since manmade nitrogen fertilizer production consumes 1–2% of global energy and generates about 3% of total global CO_2 emissions (Bourzac, 2017). The artificial nitrogen fixation process uses metal catalysts and the Haber–Bosch process to produce ammonia fertilizer from atmospheric nitrogen at high temperatures and pressures. On the other hand, nature uses the nitrogenase enzyme to (easily) fix atmospheric nitrogen into ammonia. Quantum computers might be employed for molecular simulation to understand the details of the nitrogenase mechanism to help design less energy-intensive industrial processes for synthesizing nitrogen fertilizers. Simulating the basics of nitrogen fixation may be possible with only a few hundred qubits, but also may require certain assumptions about error correction (Reiher *et al.*, 2017).

10.2.4.2 *Molecular and materials simulation*

Understanding the nitrogen fixation process is a pressing problem, but the promise of quantum computing more generally is that a new computational platform may become available for simulating every kind of molecule and materials. As Feynman envisioned universal quantum computers that simulate the world of quantum mechanical physics, Laplace dreamed of a probabilistic system which could instantiate Newtonian mechanics and "comprehend all the forces by which nature is animated" (Laplace, 1812). Laplace was instrumental in developing the probability theories that enable modern statistics.

The method used in classical computing systems for molecular simulation is called molecular dynamics (MD). MD is a computer simulation method for studying the physical movements of molecular systems in which the atoms and molecules are allowed to interact for a fixed period of time, which provides a view of the dynamic evolution of the system. However, since molecular systems typically consist of a vast number of particles, it is only possible to simulate basic configurations for short amounts of time in classical computers.

In molecular simulation, the first task of quantum computation is recapitulating already-known solutions for simple molecules such as the hydrogen molecule. In one example, research has extracted both the ground state and excited states for the H2 molecule by computing the molecular spectra on a quantum processor (Colless *et al.*, 2018). The limitations of classical computing systems make it difficult to solve problems involving even a few atoms, but near-term NISQ devices may be able to address some of these challenges. Quantum simulation initially focused on simple molecules (hydrogen and helium), but is now being used to tackle more complicated atomic configurations as well. Other research demonstrates the experimental optimization of Hamiltonian problems with up to 6-qubits and more than one hundred Pauli terms, determining the ground-state energy for molecules of increasing size, up to BeH2 (Beryllium hydride, a medium-sized molecule and alkaline earth hydride that is commonly used in rocket fuel) (Kandala *et al.*, 2017).

Considering quantum simulation as an energy problem, the idea is to estimate a chemical or other physical system's lowest energy state or ground state. For example, one research question is how to compute the ground-state energies of molecular hydrogen as a function of the inter-atomic distance between nuclei. A VQE is employed to use a Hamiltonian to test how electrons optimize their molecular arrangements to minimize binding energy (the energy required to remove an electron from an atom). Quantum chemistry problems may also involve trotterization (meaning approximating a continuous Hamiltonian by the product of a large number of small discrete rotations on 1 or 2 qubits at a time). The overall approach is first replicating the known ground states of simple molecules, and then having validated this with the quantum simulator, targeting more complicated molecules whose behavior is unknown.

Small molecule simulations, and also magnetics, are important early application areas for quantum simulation. Magnetics are implicated in the phase transitions of materials, including in spin glass problems. Understanding the magnetic phases and their transitions in quantum mechanical systems is an important objective. These kinds of systems have correlated electrons and are difficult to simulate classically. Research has developed a programmable quantum spin glass simulator to understand more about the phase transitions of complex magnetic systems (Harris *et al.*, 2018). A lattice is tuned to vary the effective transverse magnetic field, and reveals phase transition details between a paramagnetic, an ordered antiferromagnetic, and a spin glass phase. An experimental realization of a quantum simulation of interacting Ising spins on 3D cubic lattices is demonstrated. The results compare well to existing theories for this kind of spin glass problem, thus validating the approach of the quantum simulation of spin glass problems in materials physics.

10.2.4.3 *Quantum brain simulation: The final frontier*

In the longer-term, brain simulation could be a marquis application for quantum simulation. The brain is one of the most complex systems known, and the obvious goal would be to use quantum computing to simulate the brain. Whereas researchers in physics and chemistry are starting to execute simulation projects on quantum computers, there is comparatively little activity in computational neuroscience. The brain may be too complex for serious consideration of quantum simulation opportunities even with near-term NISQ devices.

Whole-brain emulation and computational neuroscience continues to advance, mainly designing algorithms for use in high-performance computing environments, but these algorithms are constrained by the speed and power consumption problems of conventional supercomputers. Teams hope that the exascale computing era (expected in the early 2020s) may offer new simulation possibilities. The important skill being developed is the ability to design efficient algorithms given the computational complexity limitations of available computing resources. One short-term goal of computational neuroscience is to advance the knowledge of neural processing in the brain, particularly regarding epilepsy and neurodegenerative disorders such as Alzheimer's disease and Parkinson's disease. A recent state-of-the-art method indicating the status of the field announces a computer that is able to simulate 80,000 neurons and 0.3 billion synapses (van Albada *et al.*, 2018) (a substantial achievement, yet compares to 100 billion total neurons in the brain).

Neuronal simulation is similar to the nitrogen fixation problem, in that nature apparently performs tasks much more efficiently than manmade methods. In the case of the brain, there is a huge gap between the energy consumption of the brain and the energy used by supercomputers attempting to simulate the brain. The possibility of formulating neural simulation problems as energy problems suggests that progress might be made towards solving them with quantum computing. This is because in quantum computation, a primary focus is on finding the lowest energy state configurations of quantum mechanical systems.

Nature easily uses the nitrogenase enzyme to fix atmospheric nitrogen into ammonia, and likewise operates 100 billion neurons in the human brain. There is little understanding of the detailed operation of neurons. Each neuron is thought to have 100,000 inputs and performs a variety of functions that result in a spike to stimulate downstream activity. A modeling approach that is adequate to the complexity of the domain might involve a compartmental model. Such a model might have a thousand separate compartments with tens of thousands of differential equations. The differential equations would execute the numerical integration of thousands of nonlinear time steps, and summarize the activity to count the voltage spikes (threshold signals) that emanate from the neuron. A simplified version of this idea is proposed for modeling the pyramidal cell (one of the largest neurons) in a dendritic tree model abstracted into a two-layer neural network (Poirazi *et al.*, 2003). A greater realization of such methods might be possible with quantum computation and human brain/computer interfaces (Martins *et al.*, 2019). In classical computing, as much as a 3,072 channel (electrode) brain–machine interface platform has been proposed (Musk *et al.*, 2019).

10.2.5 *Examples of quantum machine learning*

Quantum machine learning is an interdisciplinary field comprising quantum computing and machine learning. This could involve both the implementation of machine learning in quantum computers (e.g. a quantum computer analyzing classical data with machine learning methods), and using machine learning to build and evaluate quantum computers (Biamonte *et al.*, 2017). Quantum machine learning may also be used to refer to the study of quantum data with machine learning techniques, for example, to analyze the large volumes of data generated in experimental particle physics (Radovic *et al.*, 2018).

The two main applications in quantum computing, optimization and simulation, are conducive to deployment in quantum machine learning. Machine learning techniques excel at pattern recognition (implicated in both optimization and simulation) and have led to the widespread adoption of statistical models in computing. Machine learning models are probabilistic and statistical, and statistical distributions are likewise a native feature of quantum computing (per quantum statistics). The statistical distributions available on quantum processors (quantum statistical distributions) are likely to be a superset of those available classically. Hence, it is possible that quantum computers may outperform classical computers in machine learning tasks.

10.2.5.1 *Quantum annealing for image sorting*

Image sorting is one of the earliest use cases of quantum computing in machine learning. Google reported using the D-Wave quantum annealing machine for the binary classification of images (Neven *et al.*, 2009). The work includes numerical studies indicating that the quantum annealing method was able to compete successfully with an industry standard, Adaptive Boosting (AdaBoost, a known meta-algorithm for boosting the performance of machine learning algorithms). Google's quantum analog for boosting performance in quantum machine learning processes is called QBoost (QBoost is a large-scale classifier training with adiabatic quantum optimization) (Neven *et al.*, 2012). Quantum Monte Carlo simulations are conducted as evidence of the quantum adiabatic algorithm's ability to handle a generic training problem efficiently.

10.2.5.2 *Vanishing gradients persist in quantum neural networks*

Since the same kinds of problems are being addressed with the same kinds of methods, whether deployed in classical computers, hybrid quantum–classical, or quantum computers, some of the same kinds of challenges continue to emerge. For example, the vanishing gradients problem arises in quantum machine learning just as in classical machine learning. Barren plateaus appear in quantum neural network (QNN) training landscapes (analogous to the vanishing gradients problem) (McClean *et al.*, 2018). Vanishing gradient is the problem where a backpropagated gradient (solution slope) can quickly grow or shrink with each time step, scaling exponentially over many iterations such that it either explodes or vanishes (LeCun *et al.*, 2015).

Technically, the vanishing gradients problem refers to a failure of the algebra used in the problem solution to converge on an answer. Vanishing gradients are a known problem in classical neural networks with established workarounds. In the quantum domain, the vanishing gradients problem has surfaced in contemplating the nitrogen fixation problem. VQEs can be used to drastically reduce the problem, but it is not clear if the algorithms will converge to a solution because the variational algorithms have the same vanishing gradient problem seen in classical machine learning. It could be that a different approach is needed, which is not surprising since the same method has been reinstantiated in the quantum environment. The point is that implementing quantum machine learning is not likely going to be as straightforward as simply training a near-term NISQ device with a quantum circuit that instantiates a classical optimization algorithm.

10.2.5.3 *Supervised quantum machine learning*

A comprehensive system for developing QNNs that can represent labeled data and be trained through supervised learning is proposed (Farhi & Nevin, 2018). The model is developed using the standard dataset of labeled images of handwritten digits (the MNIST database of images of 0s and 1s). For training, the classical data are converted to a quantum format consisting of n-bit strings with binary labels. The quantum input state is an n-bit computational basis state corresponding to a sample string. The input data can be represented as quantum superpositions of computational basis states corresponding to the different label values. The quantum circuit consists of a sequence of parameter-dependent unitary transformations which act on the quantum input state. The measured output is the QNN's predictor of the binary label of the input state. For binary classification, a single Pauli operator is measured on a designated readout qubit. The work uses the classical simulation of small quantum systems to confirm the results and fine tune the model. Through classical simulation, it is demonstrated that the system finds parameters that allow the QNN to learn to correctly distinguish between the images of 0s and 1s. The QNN model could be run on a near-term gate model quantum computer to test a wider set of applications.

10.2.5.4 *Unsupervised quantum machine learning*

Another work in quantum machine learning demonstrates unsupervised machine learning on a hybrid quantum computer (a linked quantum computer and classical computer) to solve a clustering problem for recommendation engines (Otterbach *et al.*, 2017). Having effective methods for unsupervised learning is a key challenge in machine learning in general. In this research, the clustering task is translated into a combinatorial optimization problem that is solved by the QAOA. The QAOA is used in conjunction with a gradient-free Bayesian optimization to train the quantum machine (a 19-qubit Rigetti processor).

10.2.5.5 *Machine learning for better near-term NISQ devices*

Not only are machine learning problems being instantiated and run on quantum computers for more expedient problem-solving, but machine learning is also being used to develop more efficient quantum computers. Quantum dynamics are simulated on classical computers to understand and confirm the operations of quantum computers, and to develop quantum algorithms for programming quantum computers. Quantum solutions are developed in hybrid ecosystems of quantum computers and classical computers that simulate quantum behavior, and are used as the input–output layers for interfacing with quantum computers. Quantum algorithms are developed on classical machines and run on quantum computers through web interfaces. Likewise, the results are returned back from quantum computers into classical computers.

A key challenge in quantum computing is having a long enough coherence time during which qubits maintain their integrity in order to successfully perform a calculation. Qubits are fragile and can quickly decohere due to environmental noise and the natural decay of quantum states. Although robust error correction may increase the coherence time of qubits in the future, these solutions are not yet available. Due to these difficulties, currently-available quantum information processors have only very few qubits available for computing, and thus, programming the quantum algorithms for their operation must be extremely efficient.

Quantum algorithm development is undertaken primarily in classical computing environments that simulate quantum environments. Even with only a few qubits available for computing in quantum information processors, simulating the quantum dynamics of the system is taxing for classical systems. Hence, a solution is proposed in using machine learning algorithms such that NISQ devices can compile and test their own quantum algorithms (Pakin & Coles, 2019). This could avoid the substantial computational overhead required to simulate quantum dynamics on classical computers. This work extends previous work using machine learning methods on classical computers to search for shortened versions of quantum programs. More efficient algorithms for near-term NISQ devices might be produced as a result of this work.

References

Amari, S.I. (2016). *Information Geometry and Its Applications*. Heidelberg, Germany: Springer.

Bernstein, E. & Vazirani, U. (1997). Quantum complexity theory. *SIAM J. Comput*. 26(5):1411–73.

Biamonte, J., Wittek, P., Pancotti, N. *et al.* (2017). Quantum machine learning. *Nature* 549:195–202.

Bourzac, K. (2017). Chemistry is quantum computing's killer app. *ACS Chem. Eng. News*. 95(43):27–31.

Buhlmann, P. & van de Geer, S. (2011). *Statistics for High-dimensional Data: Methods, Theory and Applications*. Heidelberg, Germany: Springer.

Colless, J.I., Ramasesh, V.V., Dahlen, D. *et al.* (2018). Computation of Molecular spectra on a quantum processor with an error-resilient algorithm. *Phys. Rev*. X. 8(011021).

DiVincenzo, D.P. (2000). The physical implementation of quantum computation. *Fortschritte der Physik*. 48(9–11):771–83.

Farhi, E., Goldstone, J. & Gutmann, S. (2014). A quantum approximate optimization algorithm. arXiv:1411.4028 [quant-ph].

Farhi, E. & Neven, H. (2018). Classification with quantum neural networks on near term processors. arXiv:1802.06002 [quant-ph].

Fassler, J. (2018). Hey Amazon, Kroger's new delivery partner operates almost entirely on robots. New Food Economy.

Hadfield, S., Wang, Z., O'Gorman, B. *et al.* (2019). From the quantum approximate optimization algorithm to a quantum alternating operator ansatz. *Algorithms* 12(2):34.

Harris, R., Satol, Y., Berkley, A.J. *et al.* (2018). Phase transitions in a programmable quantum spin glass simulator. *Science* 361(6398):162–5.

Hetzner, C. (2019). VW, Canadian tech company D-Wave team on quantum computing. Automotive News Canada.

Huggins, W., Patil1, P., Mitchell1, B. *et al.* (2019). Towards quantum machine learning with tensor networks. *Quantum Sci. and Technol*. 4(2).

Kandala, A., Mezzacapo, A., Temme, K. *et al.* (2017). Hardware-efficient variational quantum eigensolver for small molecules and quantum magnets. *Nature* 549:242–6.

Kipf, T.N. & Welling, M. (2017). Semi-Supervised Classification with Graph Convolutional Networks. ICLR 2017 conference paper.

Laplace, P.S. (1812). *Philosophical Essay on Probabilities*. Trans. F.W. Truscott (Théorie Analytique des Probabilités). New York, NY: John Wiley & Sons.

Lechner, W., Haukel, P. & Zoller, P. (2015). A quantum annealing architecture with all-to-all connectivity from local interactions. *Sci. Adv.* 1(9): e1500838.

LeCun, Y., Bengio, Y. & Hinton, G. (2015). Deep learning. *Nature* 521:436–44.

Lee, J.A. & Verleysen, M. (2007). *Nonlinear Dimensionality Reduction*. Heidelberg, Germany: Springer.

Lei, N. Luo, Z., Yau, S.-T. & Gu, D.X. (2018). Geometric understanding of deep learning. arXiv:1805.10451 [cs.LG].

Malago, L., Montrucchio, L. & Pistone, G. (2018). Wasserstein Riemannian geometry of Gaussian densities. *Information Geometry* 1(2):137–79.

Martins, N.R.B., Angelica, A., Chakravarthy, K. *et al.* (2019). Human brain/cloud interface. *Front. Neurosci.* 13:112.

McClean, J.R., Boixo, S., Smelyanskiy, V.N. *et al.* (2018). Barren plateaus in quantum neural network training landscapes. *Nat. Comm.* 9(4812):1–6.

Musk, E., Neurolink. (2019). An integrated brain-machine interface platform with thousands of channels. https://www.biorxiv.org/content/10.1101/703801v1.

Neven, H., Denchev, V.S., Rose, G. *et al.* (2012). QBoost: Large scale classifier training with adiabatic quantum optimization. PMLR 2012. Proceedings: Asian Conference on Machine Learning. 25:333–48.

Otterbach, J.S., Manenti, R., Alidoust, N. *et al.* (2017). Unsupervised machine learning on a hybrid quantum computer. arXiv:1712.05771 [quant-ph].

Pakin, S. & Coles, P. (2019). The problem with quantum computers. *Sci. Am.*

Peruzzo, A., McClean, J., Shadbolt, P. *et al.* (2014). A variational eigenvalue solver on a photonic quantum processor. *Nat. Comm.* 5(4213).

Poincaré, H. (1905). *Science and Hypothesis*. New York, NY: The Walter Scott Publishing Co., Ltd.

Poirazi, P., Brannon, T. & Mel, B.W. (2003). Pyramidal neuron as two-layer neural network. *Neuron* 37(6):989–99.

Radovic, A., Williams, M., Rousseau, D. *et al.* (2018). Machine learning at the energy and intensity frontiers of particle physics. *Nature* 560:41–8.

Reiher, M., Wiebe, N., Svore, K.M. *et al.* (2017). *PNAS* 114:7555–60.

Shwartz-Ziv, R. & Tishby, N. (2017). Opening the black box of deep neural networks via information. arXiv:1703.00810 [cs.LG].

van Albada, S.J., Rowley, A.G. & Senk, J. (2018). Performance comparison of the digital neuromorphic hardware SpiNNaker and the neural network simulation software NEST for a full-scale cortical microcircuit model. *Front. Neurosci.* 12:291.

Part 4

Smart Network Field Theories

Chapter 11

Model Field Theories: Neural Statistics and Spin Glass

Abstract

Smart network field theory (SNFT) is derived from statistical physics (statistical neural field theory and spin-glass models) and information theory (the anti-de Sitter space/conformal field theory, AdS/CFT, correspondence). In this chapter, statistical neural field theory and spin-glass models as two model field theories are elaborated in a step-by-step method with regard to the various physics concepts and methods they employ. The theories are exemplars of the technophysics principle of formulating systems in a way that is analytically solvable. For both field theories, first an initial approximation of the system is made (with a mean field theory and a random energy model). Then, a more detailed analytical formulation is posited that can be solved exactly (statistical neural field theory and p-spherical spin-glass theory). The approximation reflects the system at stable equilibrium, and the solvable model, the system at criticality. Energy minimization and statistical distributions are key organizing principles in both models. The field theories are selected for their comprehensive technophysics focus in complex systems such as the brain and superconducting materials, and because these physical models continue to inspire advances in machine learning. The model theories might be used as a template for formulating field theories in any domain, whether physical or algorithmic.

11.1 Summary of Statistical Neural Field Theory

The first of the two model systems is a statistical neural field theory for describing brain activity (Cowan, 2014). A field theory of large-scale brain activity is proposed based on statistical mechanics. For the basic network description in stable non-critical situations, a mean field theory is articulated with the Wilson–Cowan equations (Wilson & Cowan, 1972). To assess system criticality and phase transition, a more complicated field theory is proposed in which network behavior evolves over time, and is represented with Markov random walks and Markov random fields as a path integral (Buice & Cowan, 2007). Recent work applies statistical neural field theory in the context of deep learning models of the visual cortex (Yamins & DiCarlo, 2016). A summary of the key aspects of statistical neural field theory appears in Table 11.1.

The overall structure of the field theory is to define the dynamical system, its state transitions, how action occurs within the system and evolves, and how system criticality and phase transition arise and might be controlled. The system is formulated as a Markov process (random walks, random fields) to derive path integrals for overall system measures such as correlation (distance) functions, moment-generating (probability distribution) functions, and actions. The action (an abstract quantity describing the aggregate motion of the system such as neural firing threshold) is derived as the path integral of the Lagrangian (the overall configuration of network fields, particles, and states).

Table 11.1. Key aspects of statistical neural field theory.

1. **Statistical field theory:** A statistical field theory is needed to model the effects of correlations and fluctuations in a dynamical non-equilibrium system such as the brain, particularly in neuronal firing.
2. **System dynamics:** A field theory is specified to identify system state transitions, evolution, and trigger points (a field theory is defined as a set of functions to assess system state and criticality).
3. **Path integrals:** The theory is formulated such that path integrals can be derived for overall system measures (actions and functions corresponding to configurations of the underlying system).
4. **Simplification and renormalization:** The model of the neural system is simplified with the application of various physics methods.

With the path integral, a master evolution equation is written for the system. A single random walk is generalized to random walks in a continuum, to frame the problem as a Markov random field. A neural dynamics for the system is obtained by quantizing the neuron into three states: quiescent, active, and refractory (QAR). A simpler two-state model (quiescent and active, QA) could also be used. Simplifications are applied such as Wick's theorem, Gell-Mann's 3×3 matrices, and eigenfunction ladder operators. An algebraic version of the master evolution equation is specified, which describes the QAR state transition diagrams of the neural network.

To articulate how action occurs within the dynamical system, a neural state vector (a probability state vector) is derived. This is possible because the algebraic equation is a number density operator that can be used to count the number of neurons in the different QAR states, and the total excitation coming onto the network (with a Hamiltonian) as a network quantity. Euclidean field theory (a modern form of statistical mechanics) is applied to derive a Wiener–Feynman path integral (one that is mathematically well defined by using imaginary time (unlike the original Feynman path integral, which is not easily computable)). This allows a simplification, to write the system with only the spike Hamiltonian instead of the full Hamiltonian (the Hamiltonian indicating the spiking threshold of the system).

Having described the dynamical system and how action occurs within it, next is defining a mathematical model of system evolution, for which coherent states are used. Coherent states are functions that do not change their state under the action of the evolution of the Schrödinger wave equation. A theory of spiking action on a neural network is obtained using coherent states, applying the U(1) symmetry (rather than that of the more complicated Lie algebra topology group SU(3)). A theory of spiking action explains system evolution.

However, the theory thus described is a linearized spiking model, so to specify a nonlinear spiking model, perturbation techniques (the renormalization group) are used to define a renormalized (rescaled) action. The issue is that there are multiple time and space scales in nonlinear systems, so a renormalized action that takes into account new critical points at longer time and length scales is necessary.

The renormalized action is the same as the action in Reggeon field theory (which characterizes strong interactions in high-energy particle physics). The behavior of a Reggeon field theory system corresponds to both branching and propagation (the crucial criticality in a neural system). Such systems have a universal non-equilibrium phase transition, in a class called directed percolation. System criticality is thus articulated in terms of directed percolated (unidirectional) phase transitions. To control the system, the action of an optimal control theory (OCT) can be expressed on the dynamical system as a path integral.

11.2 Neural Statistics: System Norm and Criticality

The problems, techniques, and results of specifying the system norm and criticality in statistical neural field theory are highlighted in Table 11.2.

11.2.1 *Mean field theory describes stable equilibrium systems*

The Wilson–Cowan equations are used as a continuum approximation (a means of modeling system kinematics as a continuous mass), inspired by technophysics work examining ferromagnetic atomic interactions in physics applied to neurons (Cragg & Temperley, 1954). The background idea is the Ising model of ferromagnetism (and phase transition). The Ising

Table 11.2. Statistical neural field theory: System norm and criticality.

Problem/Requirement	Tools/Techniques	Results
Description of the system at equilibrium	• Mean field theory • Continuum approximation	• Wilson–Cowan equations (a mean field theory)
Description of the system at criticality	• Effective field theory	• Statistical neural field theory (a statistical field theory) • A statistical field theory is necessary to describe the effects of correlations and fluctuations in dynamical non-equilibrium systems

model is mobilized in the notion of spin glasses (disordered magnets which are metastable systems with roughly half of their bonds spin-up, and half spin-down (Anderson, 1988)). Through Ising ferromagnetics incorporated as a spin-glass model, the motivation is to try to explain phase transition (such as neuronal firing) and to examine energy minimization problems (with the spin-glass model as a physical system in which energy can be minimized).

The Wilson–Cowan equations (for approximating continuous system kinematics) can be applied with symmetric or anti-symmetric weights in the formulation, which has application implications. In the neural statistical field theory, anti-symmetric weights are used to explain neuronal signaling, by constructing a theory of coupled oscillators (excitatory and inhibitory oscillators).

In machine learning, on the other hand, symmetric weights are used to build a mechanistic system for computation. The idea is using the analogy with spin glasses as a physical system in which energy can be minimized as a computational feature. Whereas asymmetric weights allow critical behavior such as neuronal firing to be assessed, symmetric weights enable a general architecture using neurons as processing units to be constructed. A symmetrically weighted Ising model is used to instantiate neurons as generic processing nodes in machine learning.

The field of machine learning was developed using the principles of spin-glass ferromagnets to create an artificial neural net, called a Hopfield network (modeled on a symmetrically weighted computational model of brain neurons) (Hopfield, 1982). A further instantiation of this kind of structure was proposed in the notion of Boltzmann machines (Ackley *et al.*, 1985). Like the Hopfield network, the Boltzmann machines define a system "energy" term as an overall network parameter and lever for calculation.

In machine learning, spin glass inspired energy functions are a key feature, and also the sigmoidal structure of neuronal processing, which is likewise defined in statistical neural field theory. The sigmoidal structure of neuronal processing is used in machine learning in the logistic regression formulation of the problem to be calculated (which results in an s-curve (sigmoid) format). In the brain, the sigmoidal structure is used to solve the credit-assignment problem for neurons (i.e. which neuron is signaling). The benefit of the sigmoidal formulation as specified in statistical neural field theory is that it is readily convertible to an algorithm for calculating system measures, whether used in analyzing the brain, or in machine learning.

11.2.2 *Statistical neural field theory describes system criticality*

The mean field theory articulated with Wilson–Cowan equations describes stable systems that are at equilibrium. However, to specify the effects of correlations and fluctuations over time in a dynamical non-equilibrium system, a more robust model is needed. In particular, a statistical field theory is used to study the brain, and to capture the state transitions and the potential movement of the system (Buice & Chow, 2013). Prior work inspired this theory development, notably research on self-organized criticality (Bak *et al.*, 1987), and the documentation of anomalously large fluctuations in spontaneous neural spiking patterns (Softky & Koch, 1993). To assess criticality and the effects of correlations and fluctuations in the non-equilibrium system, the evolution of neural activity is set up as a Markov process, and then the thermodynamic limit (continuum limit) of that expression is taken. The result of the continuum formulation is a Markov random field which can be manipulated with path integrals and other calculus-based tools.

11.3 Detailed Description of Statistical Neural Field Theory

11.3.1 *Master field equation for the neural system*

The first step in elaborating the neural statistical field theory is writing a master field equation for the neural system, as discussed in Table 11.3.

11.3.1.1 *Master evolution equation for a Markov random walk*

The first step is to obtain a random walk, by writing a master equation with neurons on a *d*-dimensional lattice (hypercube). The master equation is an evolution equation that considers all possible state transitions of the system. Structured as an evolution equation, the diffusion limit can be

Table 11.3. Obtaining a master field equation for the neural system.

Problem/Requirement

Write a system master equation considering all possible state transitions

Tools/Techniques

Brownian motion, diffusion limit, path integral

Results

System master equation written in the form of a Markov random walk

Problem/Requirement

Formulate the system as a Markov process

Tools/Techniques

Path integrals, Green's function (Gaussian propagator), Wick's theorem

Take the diffusion limit of the Markov random field to derive path integrals

Results

Path integrals derived as correlation functions, moment-generating function (an alternative probability distribution for the system), and a system action

The action (the path integral of the Lagrangian (the overall configuration of states)) is used to construct the moment-generating function (probability distribution) that describes the statistics of the network

taken, with the spacing between the neurons going to zero, and the firing time going to zero. The diffusion approximation of the random walk is obtained (diffusion is relevant as the physical observation of the Brownian motion of molecules in a random walk).

Wiener (1958) introduced the mathematical ideas of Brownian motion and the diffusion limit (although Einstein (1906) was first to study these problems in physics, and Bachelier in the context of finance (Bachelier, 1900)). The solution of the diffusion equation is a known quantity: a Gaussian propagator or a Green's function (a path integral that can be used to solve a whole class of differential equations). Wiener expressed the solution of the diffusion equation as a path integral, from which the action (an abstract quantity describing the overall motion of a system) can be derived. The path integral is structurally similar to a Hamiltonian (the total energy in a system). Hence, the same kind of mathematics is used to express the path integral as an overall measure of the system from random walks.

11.3.1.2 *Correlation functions and moment-generating function*

An important insight is that correlation functions (measures of distance in the network) can be expressed as the sum over all paths of the system, as path integrals. A moment-generating function (probability distribution) can be introduced as the path integral associated with the system. A moment-generating function is an alternative statistical means of specifying a probability distribution (in contrast to probability density functions and cumulative distribution functions), based on computing the moments in the distribution. Cumulants (another set of quantities which provide an alternative to the moments of the distribution) can be obtained the same way. These processes are well known. The correlation functions that influence system criticality are expressed in a standard way, through a moment-generating function (that is a path integral) (Ginzburg & Sompolinsky, 1994).

In Gaussian Markov random walks, the Gaussian propagator or Green's function solution to the differential equations can be used to collapse the function into a two-point correlation function. All of the higher moments depend on the first two moments, so the function can be collapsed under this solution method. For example, the two-endpoint moment function can be used to evaluate only the even moments, so the odd moments vanish, while preserving the general structure of the function. This is a result of Wick's theorem (a method of reducing high-order derivatives to a more-easily solvable combinatorics problem). With Wick's theorem, it is only necessary to compute two-point correlation functions (assuming the one-point function is zero). The result is that a usable action is obtained from the two-point correlation function, which is equal to a Gaussian propagator or the Green's function. Wick also shows that if imaginary time is used in the Schrödinger equation, it becomes the diffusion equation, further underlining the connection between Brownian motion and quantum mechanics.

11.3.2 *Markov random walk redefined as Markov random field*

The next step is expanding the Markov random walk to a Markov random field as elaborated in Table 11.4.

Table 11.4. Expanding the Markov random walk to a Markov random field.

Problem/Requirement

Generalize a single random walk to random walks in a continuum to expand from the Markov random walk to a Markov random field

Tools/Techniques

Quantize neurons into three states (QAR)

Results

Moment-generating function is rewritten for Markov random fields

Generalize the action over the network with Green's function (path integral)

Master equation rewritten as neural master equation, comprising all system states, for the Markov random field

Problem/Requirement

A model of state transitions

Tools/Techniques

Gell-Mann 3×3 matrices, ladder operators, algebraic simplification

Results

Algebraic simplification of the master equation and solution of the master equation expressed as a path integral of the Markov random field

Number density operator to count: (1) the number of neurons in the QAR states; (2) total network excitation

State diagram of the network (with neural state vector, probability state vector, neural Hamiltonian (counting all possible state transitions))

11.3.2.1 *Neural dynamics: Random walk as random field*

The key procedural step is generalizing from a single random walk to random walks in a continuum, to arrive at Markov random fields. A moment-generating function for Markov random fields can be written. This produces a field that varies over the entire network, and the action generalized from that is a derivative that includes both time and space coordinates. Gaussian random walks and fields were described previously in terms of their associated Green's functions. The same structure can be used to model and propose a neural dynamics. The neuron is quantized into three states: QAR. This is a three-state model, but it could also be a two-state model (QA neurons). The kinetics of the model is specified as the transitions between the different QAR states with rate functions and thresholds. A neural master equation for the Markov random field (not just the Markov random walk) is introduced to keep track of the states and transitions in the network.

11.3.2.2 *Algebraic simplification of the master field equation*

A standard mathematical basis, 3×3 matrices, can be used to model the transitions between the QAR states. Matrices for the representation of the Lie group SU(3) are a model for this. The algebra of quarks can be applied to the algebra of neurons. Raising and lowering operators (ladder operators) are obtained from Gell–Mann matrices. The ladder operators can be used to increase or decrease the eigenvalues of other operators. An algebraic version of the master equation can be written with ladder operators and eigenfunctions (solutions to the set of differential equations) to give the QAR state transition diagrams of the neural network.

To obtain the neural state vector, neurons are organized (as previously) in a d-dimensional lattice. The weighted configurations and their probabilities are summed to obtain a probability state vector in which the probabilities sum to one. The parallels between quantum mechanics and neuron mechanics can be seen, in that the mathematical formulations are the same. In the neural network normalization, the sum of the probabilities is one. In quantum mechanics, the sum of the square of the modulus of the functions is one (this is the wave equation).

The algebraic equation is a number density operator that can be used to count the number of neurons in the different QAR states in the network. Both, the number of active neurons can be counted, and the current or total excitation coming onto the network. The total excitation can be counted by weighting the number density of active states by the weighting function for each active neuron. Thus, the master equation can be written as a neural Hamiltonian, an evolution operator which counts all possible state transitions. The analog of the second quantized form of the Schrödinger equation in quantum field theory, viewed as Markov processes for neural networks, can be written. The result is that the solution of the master equation is expressed as a path integral of the Markov random field.

11.3.3 *Linear and nonlinear models of the system action*

The next step is expanding the Markov random walk to a Markov random field as described in Table 11.5.

11.3.3.1 *Wiener–Feynman path integral*

Following Wick's result that using imaginary time in the Schrödinger equation becomes the diffusion equation, it has been recognized that the Feynman path integrals in quantum field theory are Wiener integrals in imaginary time. This means that they can be evaluated, because the Wiener integral is mathematically well defined. It is possible to compute with the Wiener integral (which is a Gaussian integral), whereas the Feynman path integral is not mathematically well defined, and

Table 11.5. Linear and nonlinear models of the system action.

Problem/Requirement
A linear model of the system action
Tools/Techniques
Wiener–Feynman path integral interpreted with Euclidean field theory
Simplified spike Hamiltonian
Coherent states
Results
Rewrite moment-generating function with coherent states (obtain coherent state path integral)
Linearized action in the form of a linearized spiking model
Problem/Requirement
A nonlinear model of the system action (since in reality, firing rates are nonlinear across the neural field)
Tools/Techniques
Standard perturbation (renormalization group)
Results
Nonlinearized action in the form of a renormalized action
System dynamics in the form of a reaction–diffusion system to be used in determining system criticality

difficult to compute. Since the Feynman path integral cannot be calculated using imaginary time, a discipline has grown up to address this, called Euclidean (quantum) field theory (Guerra, 2005; Peeters & Zaaklar, 2011). Since Euclidean quantum field theory is a well-used modern form in which statistical mechanics is applied (Huang, 2013), it is used in the neural statistical field theory.

11.3.3.2 *Spike Hamiltonian reduction of full Hamiltonian*

A reduced model can be created by appeal to solid-state physics. A simpler neural Hamiltonian is generated as the spike Hamiltonian instead of the Hamiltonian of all possible state transitions. To do this, a minimal model is considered, one in which spike (activation) rates are low (the probability of a neuron emitting a spike is low). Most neurons exist in the quiescent state Q (thus, Q is close to one), which makes the reduced neural Hamiltonian easy to analyze.

11.3.3.3 *Coherent states*

The next step is deriving the moment-generating function for the field theory using coherent states. Coherent states are of interest because they are good representations to use for expressing the statistics of quantum field theory and Markov fields. Coherent states are functions that do not change their state under the action of the evolution of the wave equation. Schrödinger articulated coherent states in 1926, soon after introducing the wave equation, and they were further developed through work in quantum optics (Glauber, 1962). There are different kinds of coherent states. Although it might seem that the appropriate coherent states for the neural field theory would be those that express the sophisticated Lie algebra SU(3) for the reduced model, the more basic coherent states introduced by Wigner with U(1) symmetry are most effective. Using U(1) symmetry, it is possible to define an accurate theory of spiking action on a neural network.

11.3.3.4 *Neural action: Linearized coherent spiking model*

With the spike Hamiltonian and the coherent states, the moment-generating function can be rewritten. A coherent state path integral is defined for the moment-generating function. The neural action (the overall motion of the system) is obtained in terms of coherent states, meaning terms that are at most quadratic (Gaussian integrals) and thus easily computable. Next, the model is linearized (by finding a linear approximation that assesses the stability of equilibrium points). With a linearized model, the Green's function (a path integral) can be written exactly for the linearized spiking model. This is the target quantity, the action as a spiking model, since as a linearized spiking model, it can be calculated. The linearized network problem structure can be solved in the frequency space (Fourier space or momentum space) of the network.

11.3.3.5 *Neural action: Nonlinear renormalized action*

It is possible to use a linearized spiking model to describe the linear model. However, in more realistic neural models, the firing rate function is nonlinear across the field, so it is necessary to conduct a power series (Taylor) expansion of the firing norm or probability function. In this case, the action is no longer quadratic, so the perturbation methods of quantum field theory must be used for the calculation. The primary perturbation method used is the renormalization group (Chen *et al.*, 1996; Efrati *et al.*, 2013). The renormalization group is a mathematical apparatus that allows for the systematic investigation of changes in a system viewed at different scales.

Renormalization group techniques are applied to the action. The issue is that in a nonlinear system, there are multiple time and space scales. Thus, it is necessary to use a standard singular perturbation analysis to arrive at a renormalized action that takes into account new critical points at longer time scales and longer length scales. The action changes, and a renormalized action is obtained as the result. This introduces a diffusion term since on longer time scales, reactions resemble reaction–diffusion networks. This is the action that is required to obtain finite results from a nonlinear system. As time goes to infinity, the propagator approximates the propagator of a Brownian motion, and the diffusion limit of a random walk. The overall result is that at long timescales, the neural network dynamics looks like those of a reaction–diffusion system (which is important in determining system criticality).

11.3.4 *System criticality*

The next step is elaborating system criticality as highlighted in Table 11.6, with Reggeon field theory and directed percolated phase transition.

11.3.4.1 *Reggeon field theory*

The renormalized action is the same as the action in Reggeon field theory. Reggeon field theory (1975) (Abarbanel *et al.*, 1975) is an extension of Regge theory, articulated by Regge (1959) and applied by Gribov (1960s) in high-energy particle physics, in the Regge theory of strong interactions (Gribov, 2003). The behavior of a Reggeon field theory system corresponds to both branching and propagation (the important criticality in a neural system). Such systems have a universal non-equilibrium phase transition, in a class called directed percolation.

11.3.4.2 *Directed percolation phase transition*

Directed percolation is unidirectional percolation through a graph or network. A universal phase transition in nonlinear neural networks is a

Table 11.6. Statistical neural field theory system criticality.

Problem/Requirement	
	A model to assess system criticality
Tools/Techniques	
	Reggeon field theory
Results	
	Renormalized action as a Reggeon field theory action
	Reggeon field theory system (with directed percolated phase transition)
Problem/Requirement	
	A general model to detect phase transition
Tools/Techniques	
	Directed percolation phase transition
Results	
	A non-equilibrium phase transition model that adequately describes neural system criticality

directed percolated phase transition (Hinrichsen, 2009). Directed percolated phase transitions are non-equilibrium phase transitions, a universality class of directed percolation, which plays a similar role in non-equilibrium systems as the Ising model in equilibrium statistical physics (i.e. serving as a classification model for system energy and interactions) (Henkel *et al.*, 2008). When balanced or equilibrium states of a neural network are destabilized, a directed percolated phase transition can arise. Away from critical points (in which the system is stable and has low firing rates), a mean field theory such as Wilson–Cowan equations adequately describes the action and the system behavior. However, a different explanatory mechanism such as the directed percolated phase transition is needed to describe what happens in moments of system criticality and phase transition.

The emergence condition for a directed percolated phase transition is analogous to the Ginzburg condition in superconductivity (a boundary condition regarding the thickness of the superconducting plate) (Ginzburg, 1955). The phase transition occurs at a threshold that is the combination of the level of excitation in the network and the diffusion spread (the length of the diffusion spread of the activity). An upper critical dimension is obtained, just as in superconductivity. Depending on the space parameters of the network, there could be critical branching, or branching plus aggregation, which is directed percolation. Branching is avalanches of spiking neurons, and the number of avalanches scales as a power law. Thus, power laws can be seen in the system dynamics. The effects of all the nodes in the network are captured in the directed percolated phase transition formulation and provide the statistical activity (Brownian motion), which articulates system criticality.

11.3.5 *Optimal control theory*

The final step in the statistical neural field theory is deriving an OCT for managing the system. OCT is a set of mathematical optimization methods that can be used for the purpose of managing physical systems with quantitative structures, in the discipline of control engineering (Swan, 1984). The statistical neural field theory derived here can be extended into an optimal control model. In particular, the action of an OCT can be expressed as a path integral on the dynamical system (Kappen, 2005). A control theory can be externally applied as an action on the system, just as any intrinsically arising action, such as a signaling cascade, is an action on the system. The orchestration mechanism is the path integral of the action.

In the brain's neural network, a measure such as synaptic plasticity could be added to obtain a modified action as a feedback and control mechanism (Sjostrom *et al.*, 2001). This is a way that self-organized criticality (Bak *et al.*, 1987) could be embodied in the management of critical moments in the network. The goal could be to produce a self-tuning system that is orchestrated by the synaptic plasticity operator. The system could self-adjust so it reaches (or avoids) criticality on shorter timescales with synaptic plasticity as an operator that either facilitates or depresses the action. With an operator, as opposed to a constant function, the Hamiltonian would be more complex, so the underlying action would change, but not the renormalized action. The directed percolated phase transition structure could likely persist and be used as a metric for optimal system control.

11.4 Summary of the Spin-Glass Model

The second model field theory system incorporated in the smart network field theory (SNFT) work is the spin-glass model (which treats disordered ferromagnets). The key aspects are summarized in Table 11.7.

In the spin-glass model, the problem at hand is formulated as an energy function, towards the resolution of optimization problems such as protein-folding and deep learning networks. A random energy model is used to obtain a general characterization of the energy landscape. Then, a spin-glass model, as a more complicated extension of the random energy model, is applied to derive an exact solution for the system.

A random energy model is used to make random (Gaussian) guesses about the probability distribution of the overall energy available in the system. However, since these are only random guesses, the system might not ever produce order, meaning converge on a solution. In protein folding, this gives rise to Levinthal's paradox (Levinthal, 1969). The paradox is that, on the one hand, the mechanism for protein folding is unknown

Table 11.7. Key aspects of the spin-glass model.

1. **Theory formulation:** The theory is formulated as an energy function based on the Ising model of ferromagnetic interactions.
2. **Theory approximation:** A random energy model is used to approximate the total energy in the system.
3. **Theory calculation (exact):** A spin-glass model is used to solve an energy landscape convergence problem. Flat glassy landscapes are overcome by producing a glass transition, a rugged, convex funnel in the energy landscape that leads to a solution.
4. **System criticality:** The system is analyzed with energetics (the information-theoretic trade-off between energy and entropy). Phase transition occurs at a critical temperature, at the moment when the system entropy and energy converge.
5. **System management:** System performance is optimized by minimizing loss and cross entropy, modeled as an energy function convergence or glass transition.

and seems to be random, but on the other hand, proteins fold in a matter of nanoseconds, so cannot possibly be folding randomly.

To direct the energy landscape to a solution (such as obtaining a folded protein or an image recognition in a machine learning system), a more sophisticated spin-glass model is used. A spin glass is a disordered magnet that is a metastable system with about half of its molecular bonds oriented in the direction of spin-up, and the other half in the direction of spin-down (Anderson, 1988). The computational benefit of using a spin-glass model is that the spins can be relaxed into real numbers, so that the system can be solved analytically. This means that the random energy function can be instantiated as a Hamiltonian (a stratified function that is a weighted probabilistic sum of the total energy in the system) that can be solved, usually as an energy minimization problem. The related analog in statistical physics is the minimization of free energy (Merhav, 2010).

The objective of the spin-glass formulation is to have the energy landscape converge on a solution, such as producing a folded protein, or an error-minimized accurately classified set of test data in machine learning. Spin-glass models are applied to neural networks in the brain (Amit *et al.*, 1985) and deep learning systems. The aim is optimization, creating efficient loss functions that reduce combinatorial complexity and minimize cross entropy. Loss optimization is formulated as an energy function with a Hamiltonian. The overall system energy is calculated in the ground state and in subsequent states of evolution to find the glass transition, which is interpreted as a solution to the loss optimization problem.

11.5 Spin-Glass Model: System Norm and Criticality

Just as the Wilson–Cowan equations provide a good general characterization of a biological neural network through the mean field theory approach, the random energy model fulfills a similar function for the spin-glass model. The random energy model provides a basic description of the system, whereas the spin-glass model is a more complicated extension of the random energy model which can be solved exactly. The random energy model is a toy model of a system with quenched disorder. The random energy model makes random (Gaussian) guesses of the probability distribution of the total energy in the system. A summary of the problem, techniques, and results of specifying the system norm and criticality in the spin-glass model is summarized in Table 11.8.

In statistical physics, the random energy model is applied as a basic model for analyzing systems with quenched disorder (disorder is frozen (quenched) into the system). The random energy model is introduced as a possible model for glasses since glasses are amorphous (not regularly ordered) systems. However, the limitation of the random energy model is that it does not have an organizing principle, such as a Hamiltonian, to identify when and where the system might converge. To obtain a more precise measure, a Hamiltonian can be used to sum the weighted energy probability of each node's state. The total energy in the system can be

Table 11.8. Spin-glass model: System norm and criticality.

Problem/Requirement	Tools/Techniques	Results
Description of the system at equilibrium	• General characterization • Random (Gaussian) guesses at total system energy	• Random energy model
Description of the system at criticality	• Effective field theory	• *p*-Spherical spin-glass model

calculated, along with local minima and maxima, which indicate the system's critical points.

11.6 Detailed Description of the Spin-Glass Model

11.6.1 *Spin glasses*

Spin glasses are magnetic systems with quenched disorder, meaning an ongoing competition between ferromagnetic and antiferromagnetic interactions. Classically, a glass is made by applying fire to sand. Most liquids freeze when they are cooled (e.g. water). However, when a liquid is supercooled, it becomes a glass. This means that it gets viscous and only becomes amorphously solid (even if appearing fully solid in everyday use). It is thought that all liquids can be made into glasses, if they are supercooled quickly enough.

11.6.1.1 *The glass transition*

The transition to glass is not a normal and complete phase transition (as in the case of particles to a gaseous state) because the atoms are held suspended, or quenched. In the transition to glass, the arrangement of atoms is amorphous (i.e. ill-defined), but not completely random. Different cooling rates and annealing techniques produce different glassy states (glass becoming either brittle or stiff). Energetics (the relation of energy and entropy within a system) is implicated in bringing about the glass transition.

11.6.1.2 *Overcoming the glass transition problem*

The glass transition problem is explaining how a system that seems to be stuck in a flat and persistent glassy state can enter one in which the energy landscape funnels down to converge on a critical point in order to do useful work, folding a protein, for example. Levinthal's paradox arises in this domain, highlighting that although proteins appear to be folding randomly, this cannot be the case because they fold in nanoseconds, whereas trying every possible random permutation would take longer than the known time of the universe (Mayor *et al.*, 2000).

The first class of solutions suggested for addressing the glass transition remains within the physical chemistry domain in which the problem occurs, and proposes that there could be some sort of natural pattern recognition that occurs at very low temperatures. While this could account for some observed supercooling phenomena, it does not explain protein folding.

A second class of solutions is posited from the information-theoretic domain of energy and entropy, especially the idea of modeling the system as a spin glass. The notion is that the physical glass transition might be avoided entirely via energetics (with an information-theoretic model of the entropy in the system). The premise is that there is a trade-off between energy and entropy. To the extent that systems are designed with good energetics (appropriate information-theoretic use of energy and entropy), perhaps the flat glassy state can be overcome, with instead, the system ending up in a funneled convex energy landscape in which solutions converge. Levinthal's paradox and the situation of flat glassy surfaces and vanishing gradients can be resolved with a funneled energy landscape with rugged convexity, which can be understood more technically as a trade-off between system energy and entropy. The information-theoretic formulation of the glass transition is instantiated in the *p*-spherical spin-glass model.

11.6.2 *Advanced model: p-Spherical spin glass*

A conventional spin glass is a disordered magnet that is a metastable system in which roughly half of its molecular bonds are spin-up, and half spin-down (Anderson, 1988). Spin refers to the magnetic moment of an atomic nucleus that arises from the spin of the protons and neutrons. The term glass comes from an analogy between the magnetic disorder of atomic spins in the spin-glass concept, and the physical positional disorder of atomic bonds in a conventional glass. A conventional glass is an amorphous solid in which the atomic bond structure is highly irregular, as compared with a crystal, for example, which has a uniform pattern of atomic bonds. The system is called a *p*-spherical spin glass because in computing the system, the spins, which usually have spin-up or spin-down values in a quantum system, are relaxed into real numbers, and a spherical constraint is applied so that they sum to one (Barrat, 1997). These values are the spherical coordinates of a point P on a higher-dimensional spherical surface, hence the name *p*-spherical spin glass. The reason to use a spin-glass model is because the model is known and analytically solvable.

From an energy–entropy trade-off perspective, the idea is that there is a system that is being described with an energy function (a Hamiltonian), with some up-spins and some down-spins. The system is in the lowest energy state when as many as possible up and down spins are paired (e.g. neutralized). A spin-glass model is metastable at best, in that there is no one final solution. The system is forever "frustrated" with constraints that cannot be satisfied due to the structure of its molecular bonds. The properties of a spin glass are analogous to those of a semiconductor, in that the metastability makes them suitable objects for manipulation in computational (electromagnetic) systems. An important innovation for computational models of learning is that spin glasses (as an Ising model of ferromagnetism) are a physical system in which energy can be minimized as a computational feature (Hopfield, 1982).

11.6.2.1 *Computational spin-glass model*

The spin glass is a disordered ferromagnet, or orientational glass, in which the energy landscape can be directed and funneled. A spin glass can be created by articulating a ground state for the random energy model. The system energy is random, except for one state, the ground state, which is an attractor state. Specifying the energy of the ground state causes the flat glassy surface to disappear and descend into an energy funnel, similar to that of a hurricane (Garstecki *et al.*, 1999). A computational solution is thereby obtained.

11.6.2.2 *The energy landscape theory of protein folding*

In the context of protein folding, one method for minimizing the frustration in the system is by providing connections, or structural contacts, between regions in the protein (Hori *et al.*, 2009). Attempting to induce a polymer to fold without any connections results in a flat energy landscape (and never folds, per Levinthal's paradox). However, adding connections to the polymers allows the system to converge. As more and more connections are added to the protein, the energy landscape becomes increasingly funneled and rugged, and converges. The system becomes convex, and produces an energy landscape that descends; a rugged and convex spin-glass energy landscape. Further, using a spin-glass version of the random energy model allows the foldability of a protein to be computed (measured information-theoretically) (Buchler, 1999).

11.6.3 *Applications of the spin-glass model: Loss optimization*

Many systems have problems that can be structured in the form of an energy landscape that converges (i.e. an energy funnel), and these can be evaluated with the spin-glass model. Continuing in the biological domain, in addition to proteins, spin-glass models have been applied to other biomolecule conformations and interactions (Ferreiro *et al.*, 2014), and crystal nucleation (Schmelzer & Tropin, 2018). In the brain, the spin-glass model might explain the collective behavior of neural networks. Specifically, the statistical mechanics of infinite-range Ising spin-glass Hamiltonians provides one explanatory model for collective neural behavior (that is methodologically similar to the statistical neural field theory) (Amit *et al.*, 1985). Relating spin-glass models to neural networks in the brain suggests their further applicability to neural networks in computational models such as deep learning systems.

Spin-glass models are typically applied to deep learning systems for optimization, in order to create efficient loss functions that reduce combinatorial complexity. The network can run more expediently if it can distinguish relevant features quickly instead of trying all possible permutations. The backpropagation of errors method is a key advance in loss optimization techniques for deep learning networks and continues to be a focal point for research in the field (Rumelhart *et al.*, 1986). Formulating loss optimization as an energy function with a Hamiltonian is a standard technique used in deep learning systems.

One technophysics method that applies spin-glass models to deep learning uses ideas related to the complexity of spherical spin-glass models (Auffinger *et al.*, 2012). The research formulates an identity between the critical values of random Hamiltonians, and the eigenvalues of random matrix ensembles (Gaussian ensembles). The identity is then used to calculate the ground state energy of the system, and the subsequent levels of energy in the system. From the energy states of the system, the minimal energy landscape is determined, in a structure that is consistent with the transition from a glass to a spin-glass system. The formulation is applied to loss optimization in deep learning networks. In an empirical demonstration, the lowest critical values of the Hamiltonians formed a layered structure and were located in a well-defined band that was lower-bounded by the global minimum of the system (Choromanska *et al.*, 2015). The result was that the loss function of the neural network displayed a similar landscape to that of the Hamiltonian in a spin-glass model (i.e. funneled and directable).

The spin-glass model has also been used for loss optimization in deep learning in other prominent recent advances such as dark knowledge and adversarial networks. Dark knowledge is a compositing technique in which the predictions of several model runs are compressed into a single model to generate more efficient results (Hinton *et al.*, 2014). In spin-glass information-theoretic terms, the dark knowledge model continues to have the same entropy as network layers are added and compressed, but the loss function keeps improving, decreasing the error rate and delivering better results. The same structural point is true with adversarial networks. Adversarial networks describe a self-play method in which there two networks. One network, the adversary network, generates false data, and another unrelated network, the discriminator network, attempts to learn whether the data are false, not by changing the structure of the neural network, but by manipulating the convergence efficiency of the loss function (Creswell *et al.*, 2018).

11.6.3.1 *Energy–entropy trade-off within a system*

Energetics refers to the connection between energy and entropy within a system. Energy and entropy are related in the canonical phase transitions between a solid, liquid, and gas phase based on temperature and pressure, triggered by the critical temperature (T_c) (Mezard & Montanari, 2009).

In the energy–entropy trade-off, phase transition occurs at the critical temperature at which the energy and the entropy converge, when the entropy has collapsed to be the same as (converge with) the free energy in the system. The free energy and the entropy become equal. An intuitive interpretation of this formulation is that when system entropy (order decay) and energy are equal, a phase transition occurs. There are two temperature phases, one above and one below the critical temperature at which the system phase transition occurs. Below the critical temperature, the entropy vanishes and the system thermodynamics are dominated by a small set of configurations. Above the critical temperature, there is an exponentially large number of possible system configurations, and at this dispersed energy, the Boltzmann measure of entropy is roughly equally distributed between the configurations. In such a system, the "annealed" entropy density is a function of the energy density. The system is metastable at the phase transition moment because it is balanced between energy and entropy (between a low and high number of configurations). The phase transition occurs at a certain convergence of parameters.

Energy-related terms such as the Hamiltonian are important for their role in describing the trade-off between energy and entropy. Systems tend to move toward a higher entropy state over time, whether it is the universe, an office desktop, or a smart network system. The laws of thermodynamics hold in the sense that it would not be possible to transfer free energy without a corresponding energy expenditure in the form of the entropy that the system must spend. The trade-off between energy and entropy in smart network domains is a relationship between the energy and entropy terms that are maintained as the system operates, including at criticality and phase transition.

Energetic principles may also be considered in the sense that entropy measures how the energy of a system is dispersed. For example, when water is ice, the system has a high degree of order and low entropy, but when there is a phase transition and the water is water, entropy goes up because there is less order in the system (molecules float randomly and are not stored as ice crystals). The reason to think of smart network systems in terms of entropy and energy is that it casts them in information-theoretic terms, which can be useful for manipulation and calculation. Energy and entropy are related as articulated in the maximum entropy principle: for a closed system with fixed energy and entropy, at equilibrium, energy is minimized and entropy is maximized (such as a pitcher of water versus an ice block).

The energy–entropy trade-off is often involved in triggering system phase transition (Stillinger & Debenedetti, 2013). Entropy collapse (possibly in the form of a so-called Kauzmann entropy crisis) is a situation that can occur at lower temperatures within a system. The system runs out of entropy, meaning that there is a high availability of disordered energy to perform work, but it cannot be usefully applied, and a static phase transition is the result (Chamon *et al.*, 2008).

11.6.3.2 *Entropy: Coffee cup example*

Most basically, entropy is a measure of how many ways a system can be organized so that it looks the same. For example, a coffee cup with coffee and milk mixed together is a system with high entropy. There are many different fungible molecular states that produce the same overall effect. A coffee cup is at maximum entropy when the coffee and the milk are stirred in together. The more disordered the system is, the higher the entropy. On the other hand, an ordered system has low entropy. When the coffee and the milk are in two separate containers before they are mixed, the systems of the coffee and the milk are in an ordered state with low entropy. If one molecule is pulled out of the milk and placed in the coffee, it will be noticeable. Whereas in the disordered system at maximum entropy with the coffee and the milk mixed up together, it does not matter if a molecule is moved around. The mixed state is more entropic, it is more disordered than the ordered state. Overall, entropy is a way of thinking about the order or disorder of the possible arrangements of a system. A disordered system is at maximum entropy (when the coffee and milk are fully mixed together). An ordered system is at minimal entropy (when the coffee and the milk are separate).

11.6.3.3 *Cross entropy*

In machine learning, both loss optimization and cross entropy minimization are important objectives. The statistical mechanical approach to machine learning frames networks as a problem of optimizing an energy function. Loss optimization is the primary focus, but minimizing cross entropy is another important information-theoretic technique. Another way to employ the system's energy function is towards the goal of minimizing cross entropy.

Cross entropy is a measure of the efficiency of an information encoding system. For example, in an anti-collision autonomous driving system that, at the highest level, distinguishes between vehicles and non-vehicular objects, a binary yes–no suffices for information encoding. On the other hand, if a smart city traffic sensor system is identifying vehicles by make and model, more bits are required to efficiently encode the information needed to label the data. In the first example, cross entropy is minimized if the simplest binary encoding scheme is used. In the second example, slightly greater sophistication is required for cross-entropy minimization. Overall, cross entropy is an analytical technique that can be used to quantify and compare the difference between two probability distributions and their efficiency in information encoding in information-theoretic systems.

References

Abarbanel, H.D.I., Bronzan, J.D., Sugar, R.L. & White, A.R. (1975). Reggeon field theory: Formulation and use. *Phys. Rep.* 21(3):119–82.

Ackley, D.H., Hinton, G.E. & Sejnowski, T.J. (1985). A learning algorithm for boltzmann machines. *Cog. Sci.* 9:147–69.

Amit, D.J., Gutfreund, H. & Sompolinsky, H. (1985). Spin-glass models of neural networks. *Phys. Rev. A.* 32(2):1007–18.

Anderson, P.W. (1988). Spin glass I: A scaling law rescued. *Phys. Today* 41(1):9.

Auffinger, A., Ben Arous, G. & Cerny, J. (2012). Random matrices and complexity of spin glasses. *Commun. Pure Appl. Math.* 66(165):165–201.

Bachelier, L. (1900). Théorie de la speculation. *Annales Scientifiques de l'École Normale Supérieure*. 3(17):21–86.

Bak, P., Tang, C. & Wiesenfeld, K. (1987). Self-organized criticality: An explanation of 1/f noise. *Phys. Rev. Lett.* 59:381–4.

Barrat, A. (1997). The p-spin spherical spin glass model. arXiv:cond-mat/9701031 [cond-mat.dis-nn]. 1–20.

Buchler, N.E.G. (1999). Universal correlation between energy gap and foldability for the random energy model and lattice proteins. *J. Chem. Phys.* 111:6599.

Buice, M.A. & Chow, C.C. (2013). Beyond mean field theory: Statistical field theory for neural networks. *J. Stat. Mech.*P03003:1–24.

Buice, M.A. & Cowan, J.D. (2007). Field-theoretic approach to fluctuation effects in neural networks. *Phys. Rev. E* 75:051919.

Chamon, C., Cugliandolo, L.F., Fabricius, G. *et al.* (2008). From particles to spins: Eulerian formulation of supercooled liquids and glasses. *PNAS* 105(40):15263–8.

Chen, L.Y., Goldenfeld, N. & Oono, Y. (1996). Renormalization group and singular perturbations: Multiple scales, boundary layers, and reductive perturbation theory. *Phys. Rev. E* 54(1):376.

Choromanska, A., Henaff, M., Mathieu, M. *et al.* (2015). The loss surfaces of multilayer networks. *AISTATS* 38:192–204.

Cowan, J.D. (2014). Personal Account of the Development of the Field Theory of Large-Scale Brain Activity from 1945 Onward. In: Coombes *et al.* (Eds). *Neural Fields: Theory and Applications*. Heidelberg, Germany: Springer.

Cragg, B. & Temperley, H. (1954). The organization of neurones: A cooperative analogy. *Electroencephalogr. Clin. Neurophysiol.* 6:85–92.

Creswell, A., White, T., Dumoulin, V. *et al.* (2018). Generative Adversarial Networks: An Overview. *IEEE Signal Process. Mag.* 35(1):53–65.

Efrati, E., Wang, Z., Kolan, A. & Kadanoff, L.P. (2013). Real space renormalization in statistical mechanics. arXiv:1301.6323 [cond-mat.stat-mech].

Einstein, A. (1906). *Zur theorie der brownschen bewegung* (On the theory of Brownian motion). *Annalen der Physik.* 19:371–81.

Ferreiro, D.U., Komives, E.A. & Wolynes, P.G. (2014). Frustration in biomolecules. *Q. Rev. Biophys.* 47(4):285–363.

Garstecki, P., Hoang, T.X. & Cieplak, M. (1999). Energy landscapes, supergraphs, and 'folding funnels' in spin systems. *Phys. Rev. E. Stat. Phys. Plasmas Fluids Relat Interdiscip Topics* 60(3):3219–26.

Ginzburg, I. & Sompolinsky, H. (1994). Theory of correlations in stochastic neural networks. *Phys. Rev. E* 50:3171.

Ginzburg, V.L. (1955). On the theory of superconductivity. *Il Nuovo Cimento (1955–1965)*. 2(6):1234–50.

Glauber, R.J. (1962). The Quantum theory of optical coherence. *Phys. Rev.* 130:2529.

Gribov, V.N. (2003). The theory of complex angular momenta. *Gribov Lectures on Theoretical Physics*. Cambridge, UK: Cambridge University Press.

Guerra, F. (2005). Euclidean field theory. arXiv:math-ph/0510087.

Henkel, M., Hinrichsen, H. & Lubeck, S. (2008). *Non-Equilibrium Phase Transitions*. Heidelberg, Germany: Springer.

Hinrichsen, H. (2009). Observation of directed percolation: A class of nonequilibrium phase transitions. *Physics* 2:96.

Hinton, G., Vinyals, O. & Dean, J. (2014). Distilling the Knowledge in a Neural Network. NIPS 2014. Deep Learning workshop.

Hopfield, J.J. (1982). Neural networks and physical systems with emergent collective computational abilities. *Proc. Natl. Acad. Sci.* 79:2554–8.

Hori, N., Chikenji, G., Berry, R.S. & Takada, S. (2009). Folding energy landscape and network dynamics of small globular proteins. *PNAS* 106(1):73–8.

Huang, K. (2013). A critical history of renormalization. *Int. J. Mod. Phys. A* 28(29):1–27.

Kappen, H.J. (2005). Path integrals and symmetry breaking for optimal control theory. arXiv:physics/0505066 [physics.gen-ph].

Levinthal, C. (1969). How to fold graciously. *Mossbauer Spectr. Biol. Syst. Proc.* 67(41):22–6.

Mayor, U., Johnson, C.M., Daggett, V. & Fersht, A.R. (2000). Protein folding and unfolding in microseconds to nanoseconds by experiment and simulation. *PNAS* 97(25):13518–22.

Merhav, N. (2010). Statistical physics and information theory. *Foundations Trends Commun. Inf. Theory*. 6(1–2):1–212.

Mezard, M. & Montanari, A. (2009). *Information, Physics, and Computation*. Oxford, UK: Oxford University Press, pp. 93–168.

Peeters, K. & Zaaklar, M. (2011). Euclidean field theory. Lecture notes. http://maths.dur.ac.uk/users/kasper.peeters/pdf/eft.pdf.

Rumelhart, D.E., Hinton, G.E. & Williams, R.J. (1986). Learning representations by back-propagating errors. *Nature* 323:533–6.

Schmelzer, J.W.P. & Tropin, T.V. (2018). Glass transition, crystallization of glass-forming melts, and entropy. *Entropy* 20(2):1–32.

Sjostrom, P.J., Turrigiano, G.G. & Nelson, S.B. (2001). Rate, timing, and cooperativity jointly determine cortical synaptic plasticity. *Neuron* 32:1149–64.

Softky, W.R. & Koch, C. (1993). The highly irregular firing of cortical cells is inconsistent with temporal integration of random EPSPs. *J Neurosci.* 13:334–50.

Stillinger, F.H. & Debenedetti, P.G. (2013). Glass transition thermodynamics and kinetics. *Annu. Rev. Condens. Matter Phys.* 4:263–85.

Swan, G.W. (1984). *Applications of Optimal Control Theory in Biomedicine*. Boca Raton, FL: Chapman & Hall/CRC.

Wiener, N. (1958). *Nonlinear Problems in Random Theory*. Cambridge, MA: Cambridge University Press.

Wilson, H.R. & Cowan, J.D. (1972). Excitatory and inhibitory interactions in localized populations of model neurons. *Biophys. J.* 12:1–24.

Yamins, D.L.K. & DiCarlo, J.J. (2016). Using goal-driven deep learning models to understand sensory cortex. *Nat. Neurosci.* 19:356–65.

Chapter 12

Smart Network Field Theory Specification and Examples

Abstract

Smart network field theory (SNFT) is defined as a conceptual and formal tool for the characterization, monitoring, and control of smart networks. Smart networks are complex systems with thousands, millions, or billion many-particle elements, and therefore, a well-grounded analytic system-level theory, with robust foundations in physical science, is needed. The SNFT is derived from statistical physics (statistical neural field theory and spin glass models) and information theory (the anti-de Sitter space/conformal field theory, AdS/CFT, correspondence). The practical use of the SNFT is for fleet-many item orchestration and system criticality detection in smart network systems. In this chapter, SNFT is specified in terms of system structure, operation, and criticality, and potential applications are considered in blockchain and deep learning smart network systems.

12.1 Motivation for Smart Network Field Theory

Smart network field theory (SNFT) construction is motivated by the fact that many smart network technologies are effectively a black box whose operations are either unknown from the outset (deep learning networks) or becoming hidden through zero-knowledge proof technology (blockchain economic networks). Simultaneously, there is a rapid worldwide adoption of smart network technologies with unknown effects. The possible advent of quantum computing introduces even greater uncertainty. The intuition that smart networks are similar to quantum many-body systems in which interactions become too complex to model directly is used to define SNFT.

12.2 Minimal Elements of Smart Network Field Theory

The theoretical foundations of SNFT development are joined with features from the model systems (statistical neural field theory and the spin glass model) and applied to smart networks to instantiate SNFTs on a more concrete basis. Three basic elements are proposed as a minimal configuration for SNFTs (Table 12.1).

The system structure includes a definition of the particles (system units) and their interactions. The system dynamics and operation describe how the system operates and changes with time. Relevant functions include temperature terms as a snapshot of system metrics, Hamiltonian terms as a representation of the dynamics of the system, and to act as operators on the system, and scale-spanning mechanisms that indicate the optimal level for interacting with the system for certain functions. The system criticality analysis includes threshold triggers as to when the system may indicate anomalous behavior, symmetry-breaking, and phase transition, and suggested strategies for the optimal control of the system.

Table 12.1. SNFT: Minimal elements.

Element	Function
1. System structure	Particles (system units) and interactions
2. System dynamics and operation	Temperature term (system metrics) Hamiltonian term (system operator) Scale-spanning (system evolution)
3. System criticality	Threshold trigger Phase transition Optimal control mechanism

12.3 Smart Network System Definition

The first step in articulating a SNFT is to define the system by identifying the particles and their potential interactions within the system (Table 12.2). The "particles" are the units in the system that comprise the system (there could be various and multiple types). The interactions are the different ways that the particles or units might behave independently and in interactions with other particles or units.

12.4 Smart Network System Operation

The second step in articulating a SNFT is defining the parameters of the system operation. These include the relevant parameters (standard and custom) for the system operation. The basic parameters might include temperature terms, Hamiltonian terms, scale-spanning terms,

Table 12.2. SNFT: Particles and interactions.

	Field theory domain	Particles	Interactions
1.	Traditional physical system	Pollen molecules in water	Particle vibration, Brownian motion
2.	Neural network in the brain	Neurons	Signaling, firing cascades
3.	Protein folding	Nucleotides	Structural contacts
4.	Blockchain	Peer-to-peer nodes	Consensus and transaction confirmation
5.	Deep learning	Artificial neurons (perceptrons)	Error calculation and loss minimization
6.	Quantum internet	Secure keys (PKI infrastructure)	Key distribution and revocation
7.	UAV/autonomous driving	Drones/automotive vehicles	Behavioral coordination
8.	Verifiable markets	Buyers and sellers	Purchases of goods and services
9.	Programmatic trading (HFT)	Trades, trading agents	Trading, liquidity management
10.	Medical nanorobots	Nanorobots	On-demand grid formation

Table 12.3. SNFT: System operating parameters.

	System operating parameter	Function
1.	Temperature term	Macroscopic label for microscopic behavior
2.	Operators (Hamiltonian, Lagrangian) and variational principles (action, path integral)	Point value (real number) for a dynamical system configuration
3.	Scale-spanning mechanism	Indication of optimal level for system engagement, and portability across system scales

and any other aspects specialized to the system at hand (Table 12.3). The temperature term provides a descriptive summary metric at the overall systems-level. The Hamiltonian term is a point value that corresponds to an entire configuration of elements in the underlying system. For example, one use of the Hamiltonian term could be as a formalized measure of the system capacity as to what number or percent of system nodes are available to provide certain services. The scale-spanning mechanism gives an indication of the optimal level for interacting with the system for certain kinds of functions.

12.4.1 *Temperature term*

Physics terms may have two meanings in the context of SNFTs, one that is precise and analytical, and one that is conceptual and analogical. Terms such as *temperature* and *pressure* might be used in different ways in smart network systems. This could be literally as a physical quantity, conceptually as a system control lever, and analytically as a system assessment parameter.

The justification for a temperature term in smart network systems is as follows. The claim is that any kind of system that can be described with statistical physics, including smart networks, may have different kinds of probability distributions, including Maxwell–Boltzmann distributions (probability distributions of a quantity) and ensemble distributions (probability distributions over all possible states of a system). The probability distributions allow a macroscopic term to be formalized. This means that any random conserved energy phenomenon at the microscopic level of a system will have a macroscopic term capturing the aggregate activity (e.g. temperature, pressure, etc.). A temperature term occurs naturally in any system in which there is a conserved quantity, which could include energy or system state, and therefore applies to smart networks.

In the SNFT, temperature is first used literally, in the operational sense as a physical quantity that expresses hot and cold. Temperature is a proportional measure of the average kinetic energy of the random motions of the constituent particles of matter (such as atoms and molecules) in a system. Smart network systems are physical, and thus temperature as heat makes sense to infrastructure operators and any party focused on the physicality of the system, such as for design and risk management purposes. In smart networks, one form of the temperature term is a quantitative measure of the overall system, such as blockchain hash rate and deep learning system error rate.

In the SNFT, temperature is also used conceptually, as the temperature term is based on theoretical arguments from thermodynamics, kinetic theory, and quantum mechanics. A temperature or pressure term links the levels in a system between a microscopic and a macroscopic domain. Temperature is the consolidated measure that describes the movement of all of the particles in a room. Likewise, in a field theory, a temperature term is a consolidated measure that comprises all of the activity of a microscopic level at one or more tiers down in the system. The temperature term might be employed as a control lever for the system.

The third use of a temperature term in a statistical mechanical system is as a system assessment parameter. In deep learning systems, examples of such a temperature term are regularization and weight constraints in the system activation function. The sum of the weights on the node activations comprises a temperature term that can be used to assess or manage the system. Overall, the temperature term can be conceived as a valve for controlling a system-level quantity or resource, setting the scale of the "energy" or capacity in the system, which might be increased or decreased as a regulation and control mechanism. An emerging layer of network resources (such as platform entropy, algorithmic trust, error contribution, liquidity, or other information attributes) might be analogs to energy in a traditional physical system and managed similarly as levers for system-level regulation.

As a practical example, temperature and pressure terms are macroscopic measures that are chosen for being informative. For example, if the pressure rises above some specified working range or the temperature drops below it, it is an indication of a possibly dangerous malfunction of the system that must be overseen. An informative measure for blockchain systems would be early warning systems of impending events. This could include information that a 51% attack is forming, that various factors may be compromising network security, that stablecoins may need reserve rebalancing, that new payment channel factories are open, or that timing is optimal for lowest-fee transactions.

Table 12.4 sets forth system operating parameters in the form of a temperature term, a Hamiltonian term, and a system-spanning mechanism in the various smart network systems.

12.4.2 *Hamiltonian term*

The use of a temperature term highlights the theme of having higher-level levers in a system that correspond to a whole class of underlying activity. Whereas temperature and pressure are macroscopic labels for one set of microscopic phenomena, another set of levers is available to capture the dynamics of the system, in the form of operators and variational principles. Operators and variational principles are terms that track the variability of certain quantities in a system. In the traditional physics setting, there are energy, momentum, and position operators, among others. Energy operators (which measure the total amount of energy in the system over time) may be the most familiar, such as the Hamiltonian. The action is a path integral of the operator (a summation over all paths or trajectories of the system), from which the equations of motion of the system can be derived. For example, in Cowan's statistical neural field theory, the action is a path integral of the Lagrangian (the overall configuration of possible states), which is used to construct the moment-generating function, the threshold at which neural spiking (signaling) occurs in the brain.

The benefit of using operators and variational principles is that a real number can be obtained as a point value that corresponds to a specific configuration of the underlying system. Further, the functions assign a real number to a dynamical system configuration that includes spatial and

Table 12.4. Operating parameters in smart network systems.

Field theory domain	Field theory parameters: 1. Temperature term	2. Hamiltonian term	3. Scale-spanning mechanism
1. Traditional physical system	Temperature, pressure, volume, amount	Hamiltonian, Lagrangian, action, path integral	Measurement schema
2. Neural network in the brain	QAR signaling potentiation	Firing rate function, action	Reggeon field theory
3. Protein folding	Polymer folding capacity	Connections, Hamiltonian	Conformation signaling
4. Blockchain	Hash rate, confirmed block, account balance	Blockchain Hamiltonian (trust, quantity, function)	Merkle Tree (hashes), accumulators, atomic swaps
5. Deep learning	Feature abstraction, error rate	Activation (ReLU), regularization	LSTM, adversarial nets, dark knowledge
6. Quantum internet	Entanglement	Superposition, quantum-secure network	Toffoli gate quantum computing
7. UAV, autonomous vehicle network, robotic swarm	Coordination, degree of thrashing	Secure system, resource-starved system	Information sharing, group operations
8. Verifiable markets	Proof-based activity	Trust barometer	Payment channels
9. Programmatic trading (HFT)	Volatility	Liquidity	Reflexivity broadcast
10. Medical nanorobots	Lipofuscin expelled	Homeostasis Hamiltonian	Cell-organ-tissue signals

Notes: **Temperature term:** A consolidated measure comprising a microscopic level of activity at one or more tiers down in a system. **Hamiltonian term:** An operator or variational principle producing a point value of an underlying system configuration. **Scale-spanning mechanism:** An indication for spanning multiple system levels, including optimizing intervention where the function curve is the smoothest.

temporal aspects. The implication is that time, space, and geometrical complexity parameters might be varied, the effect of which can be measured. Overall, operators and variational principles might be extremely useful when applied to identify and manage critical points and phase transitions in smart network systems. For example, one way that a smart network Hamiltonian might be employed is to measure the point value of system capacity. This could be the number or percent of system nodes that are available to provide a certain service. The smart network Hamiltonian could be a quantitative measure of system capacity that could be employed natively to rebalance the system (like a stablecoin). The blockchain Hamiltonian, for example, is a concise algebraic way of writing down the master equation of the potentially thousands and millions of nodes in a blockchain network.

12.4.3 *Scale-spanning portability*

A key objective of a field theory is to be able to transition portably across system scales. Smart network technologies (such as blockchain and deep learning) are not just one microscopic and macroscopic level, but rather may extend for many levels up and down, both quantitatively and qualitatively. Two examples of multi-level quantitative spanning are as follows. First is the situation where the US Treasury officials track dollar bills at 20 different levels of details. In a blockchain system, arbitrarily many levels can be rolled up with a Merkle tree structure, as the entire Bitcoin blockchain (over 580,000 transaction blocks as of June 30, 2019) can be called with one composite value, a single hash code or Merkle root. Controlled substance pharmaceutical inventories could be similarly tracked in blockchain-based digital asset registries and aggregated at any level of detail, on a hospital, county, state, national, or international basis. It could be expected that eventually nearly all assets might be registered, tracked, and managed with blockchain-based digital inventories (Swan, 2019).

Turning to deep learning networks, there are examples of spanning both quantitative and qualitative system levels. A basic deep learning network might have 5–8 layers of processing nodes on average, and different ways of abstracting the operation of these nodes into higher-level feature sets. Convolutional networks combine simple features (a jaw line) into more complex feature sets (a face) for image recognition, and recurrent networks identify relevant sequences from data in speech recognition, using LSTM (long short-term memory) to flag what to remember and what to forget. Adversarial networks (two networks pitted against each other) (Creswell *et al.*, 2018) and dark knowledge (information compression system) (Hinton *et al.*, 2014) are examples of higher-level roll-ups that are qualitative (i.e. different in kind, not only in magnitude). Qualitative scale-spanning can also be seen in recursive deep learning, in which machine learning is applied to the improved design of machine learning systems themselves. In recursive deep learning, machine learning instead of hand-coded optimization algorithms are used (Andrychowicz *et al.*, 2016), including to find new topologies of deep learning architectures (Miikkulainen *et al.*, 2017).

Recognizing that smart networks are possibly significantly many multi-level systems, the question arises: at which optimal level should the system be engaged for specific operations? One strategy is using SNFT to produce an overall equation for the system (analogous to Cowan's system master equation and nonlinear model of the system action) and apply differential calculus to evaluate where the curve is the smoothest (Harrison, 1999). The field theory can be used to determine at which derivative or integral level in the system, the curve of a certain system function is the smoothest. For example, in another model system, financial market options trading, there are at least four system levels: price, delta, gamma, and vega. The advanced terms are all partial derivatives of price; delta is the rate of change of price, gamma is the rate of change of delta, and vega is the rate of change of gamma. In this system, one of the strongest signals can be at the vega level (this is where the curve is the smoothest), and many traders engage the market at this level in so-called vega trading (Ni *et al.*, 2008). In smart network systems, partial derivatives of the Hamiltonian term could likewise be taken as one means of determining where the curve is the smoothest.

12.5 Smart Network System Criticality

The third step in articulating a SNFT is developing an understanding of system criticality, and an optimal control theory. Relevant aspects could include threshold triggers as to when the system may indicate anomalous behavior, phase transitions, and suggested mechanisms for the optimal control of the system. A SNFT should be able to make relevant and testable predictions. An integrated discussion of SNFT systems is presented, first, outlining the particles (nodes) and their potential states, second, the system actions taken on the nodes (per a Hamiltonian-type operator), and third, the implications for system state transitions and criticality.

System criticality is a key concern in any complex system. Changes in the individual components of communications and computing networks (smart networks) may cause large-scale networks to undergo phase transition, and transform rapidly from one behavior to another. This means that a stable network can quickly become unstable if even a few components fail or act anomalously, so understanding how to model these effects is essential. SNFT employs network mathematics and complexity theory to analyze critical behavior. In particular, graph theory and matrices are employed to capture the patterns of connection and interaction within smart networks. Previous work in network complexity is used to study the interrelation of system parameters such as how congestion depends upon traffic volumes and other incident phenomena (Kelly, 2008).

12.5.1 *Particles (nodes)*

In smart networks, the *particles* are the constituent network elements. In blockchains, this is the peer-to-peer nodes, and in deep learning networks, the perceptron processing units. In blockchain networks, the core unit is the peer-to-peer node. This is the flat-hierarchy, distributed peer-to-peer nodes that offer fee-based services for activities such as transaction validation and recording (mining), and currently free services such as record-keeping (ledger-hosting). There are also other units or nodes in blockchain ecosystem such as wallets, transactions, smart contracts, digital assets, and digital asset inventories.

In deep learning networks, the constituent units are perceptron nodes. Perceptrons are modular units, organized in a graph-theoretic structure. The computation graph has nodes and edges. Input values enter the processing node from the input edges, are processed by logic in the node, and exit as an output value on the output edge, which is used as the input value for a downstream node. Thousands and millions of nodes are organized Lego-like into cascading layers of processing (5–8 layers of thousands or millions of nodes in a basic deep learning network).

12.5.2 *Node states*

The nodes, units, or particles in the system may have different possible states and values. For example, wallets in a blockchain network may have a ledger balance (e.g. an amount of money, contract status, or program memory state). Wallet states are analogous to the neuronal states in the Cowan statistical neural field theory. Smart networks are state transition machines (systems that track individual and global state changes). The state values might be binary (quantal; discrete), or continuous, depending on the modeling technique. A field-theoretic model can expand in complexity. Having continuous state values would require a more complicated configuration, and eventually introduce operators instead of merely point functions.

Nodes have states. In the statistical neural field theory system, biological neurons might be in one of two or three different states (quiescent or active (QA), or also refractory (QAR)). In blockchain systems, the same two-state structure could be used initially, to model nodes that are either broadcasting or quiescent (BQ). In deep learning networks, different node states might likewise be identified. These could include a basic measure of whether the node is processing or finished (PF), or more complicated measures such as a maximum or minimum error for the particular node, or a continuous value for error-maximizing contribution. The system may be already set up to operate by calculating the maximum error contributed by each node and a field theory operator could be a more effective or an alternative aggregator of such information. In deep learning systems, the softmax function is essentially a field-theoretic application of a renormalization function. Similar to the spin glass model's relaxation of spin values to real numbers for calculation, the softmax function exponentiates a set of numbers, and rescales them so that they sum to one, allowing a state interpretation based on probability. The general idea is that field-theoretic formulations such as the softmax function allow deep learning networks to be evaluated with a Hamiltonian-type system-wide probability measure.

12.5.3 *Node action*

The node actions of a system can be measured with a temperature term or a Hamiltonian-type term. The softmax function is one example of a Hamiltonian-type operator in a deep learning smart network which is used to measure an overall system quantity (in the case of softmax, individual processing node error contributions which consolidate into the overall error rate of the network). More generally, the idea is to identify system-level quantities that would be the target of a SNFT, and to create a Hamiltonian-type term to measure them. One of the most basic system quantities to measure could be system capacity, namely the overall capacity of nodes to provide services. For example, in a blockchain network, a Hamiltonian-type operator could be used to sum the probability of nodes that are in the broadcast state of providing or being able to provide certain services. Likewise, in a deep learning network, many perceptrons are idle and have additional processing cycle capacity available that could be devoted to additional tasks.

A temperature term could be defined as the quantitative amount of network resource availability (expressed as a probability percentage, or as abstracted into a fixed quantitative metric such as temperature, which has scale and meaning within the system). The probability of network resource availability could be calculated by the weighted probability of certain nodes being in certain states (e.g. Mining On (Y-N), Ledger-Hosting (Y-N), Credit Transfer (Y-N)). In deep learning, the same concept of aggregate node states that are measured as a global system variable could be used. The system quantities that are calculated could include error contribution, feature abstraction levels, and degree of sequentiality (in an long short-term memory recurrent neural net, LSTM RNN). SNFTs are intended as a general purpose tool that could be deployed to measure a variety of aggregate network resources. For example, there could be the concept of "Cowan for Ripple," meaning the Cowan sum of the network node probability of credit availability and overall network liquidity. Knowing at a glance that there is a significant contraction in liquidity is extremely valuable because credit contraction is a potential harbinger of financial crashes.

Smart network resources (quantities) might be modeled as binary availabilities, for example, the peer-to-peer network node status of BQ distributions (broadcast or quiescent, just like the binary neuron system quiescent or active, QA), or as continuous values. The first step is defining metrics that are easy to measure in the system, such as the total percent of nodes that are live and providing a service such as credit or image recognition. More complicated models could measure the magnitude and change in magnitude, with quantities at the smoothest part of the credit curve or feature identification curve. Multi-level systems analogous to financial markets could be implemented, for the relevant metrics equivalent to the price–delta–gamma–vega partial derivative stack in options trading (and also introducing theta as a time parameter to build in the temporal dynamics of the system). The metrics could be evaluated either as an interpretation of mathematical smoothness, or as a system-abstracted value such as temperature.

12.5.4 *State transitions*

The objective of a SNFT is to characterize, monitor, and control smart network systems. Therefore, an important goal of a field-theoretic formulation is the ability to identify system phase transitions. This means both positive and negative state transitions, and also first-order and second-order transitions (e.g. a change per a continuous parameter or a dynamical evolution of the system) (Sole *et al.*, 1996). State transitions are the threshold levels of node actions (system-level quantities measured). The point is to understand what happens when the system reaches certain threshold levels of system-wide quantities. The quantities could be measured with temperature terms, operators (Hamiltonian and Lagrangian terms), or variational parameters and system path integrals (actions). The complete analysis would capture the system-wide calculation of the probability that certain nodes are in certain states, derive an interpretation of how this impacts system criticality, and monitor and control such criticality.

The Hamiltonian-type term in the SNFT could be used to measure system quantity metrics as system-wide resource indicators, and system resource exchange metrics as system-wide economic-type indicators. There could be other classes of risk-assessment metrics. In a practical example, a SNFT could be used to understand and evaluate systemic risk and the possibility of flash crashes in traditional financial markets. Evaluating system criticality is important as this domain is undergoing a phase transition to include both human and algorithmically driven market participants. A substantial portion of worldwide equities market activity is executed through programmatic trading.

12.5.4.1 *Hidden microscopic-level system data*

Not only is a temperature term a useful macroscopic control mechanism for a system comprising an enormous volume of microscopic activity, it is necessary in systems in which microscopic-level data may not be available. Mechanisms such as SNFTs could be crucial in domains such as smart networks for network resource monitoring and control because less information is available. This includes private transactions in blockchains and the native operation of deep learning, which is a black box (operational details are unknown).

In smart networks, since less microscopic information may be available, what is needed is a meta-information overlay regarding the aggregate activity of the system. More specifically, some of the ways that less information is available in the privacy-protected computing era of smart networks include (1) confidential transactions (using zero-knowledge proof technology), (2) permissioned enterprise blockchains (limited transaction visibility to participating parties), and (3) data markets (unlocking new data silos (health information, validated social network referrals) with private transactions). The question is, how to obtain an overlay metric that provides aggregate information about the computing network? Similarly, in deep learning, the architecture, by definition, comprises hidden layers whose micro-computation is unknown at the macroscopic level. This is an important benefit of SNFT. Smart network systems require new models for their evaluation and control, which SNFT provides.

A defining feature of smart networks is that they may be simultaneously private and transparent, meaning that the whole system operates per the same parameters (transparent), but the individual microscopic level transactions (blockchain) and forward-back-propagations (deep learning) are hidden. This is another argument for why meta-level temperature terms are needed to understand overall system behavior. Accurately calculated mathematical formalisms per SNFTs are a means of deriving a well-formed temperature term for a network, especially because the microscopic-level "particle movement" could be a system parameter that is unknown.

Table 12.5 presents system criticality information for smart network systems, including threshold trigger, phase transition, and optimal control mechanism.

Table 12.5. System criticality parameters in smart network systems.

Field theory domain	Threshold trigger	Phase transition	Optimal control mechanism
1. Traditional physical system	Temperature, pressure level	Gas: ice, liquid, vapor	Critical temperature
2. Neural network in the brain	Neuronal spike	Signaling cascade	Path integral of the OCT as a system action
3. Protein folding	Rugged energy funnel	Folded protein	Polymer dynamics
4. Blockchain	Consensus, algorithmic trust, liquidity	51% attack, system fork	Early warning signal
5. Deep learning	Classification algorithm, recognition, optimization	Convergence, endless loop, vanishing gradient	Layered architecture, adversarial net, dark knowledge methods
6. Quantum internet	Stalled key issuance	Bell pair entanglement failure	Quantum error-correcting codes
7. UAV, autonomous vehicle network, robotic swarm	Collision, deadlock	Lack of coordination	Collective intelligence
8. Verifiable markets	Volume decline	Computational trust collapse	Computational proofs
9. Programmatic trading (HFT)	Circuit breakers	Flash crash	Vega hedging
10. Medical nanorobots	Disease	Pathology resolution	Therapeutic intervention

12.5.4.2 *Conclusion on SNFT analysis*

Overall, the method of elaborating a series of side-by-side comparisons of the definition, operation, and criticality of traditional and smart network systems indicates two things. First is that smart network systems can be defined in a structure that is conducive to study as SNFTs. Second is that SNFTs are shown to be a good model for studying, characterizing, monitoring, and controlling smart network systems with physics-based principles.

12.6 Applications of Smart Network Field Theories

12.6.1 *Smart network service provisioning application layers*

To concretize SNFTs, this section outlines some specific examples of the kinds of system resources that might be measured by a Hamiltonian-type term, and how system dynamics might signal system criticality and phase transition. SNFTs might be used to provision and deploy both network resources and application layers on the smart network stack. Similar to the operation of classical communications network provisioning, two of the most straightforward application layers could be instantiated as basic services and value-added services.

Nodes might provide peer-based services to other peers in the network, whether blockchain peer-to-peer nodes or deep learning perceptron nodes. Tables 12.6 and 12.7 list some of these kinds of services for blockchain networks and deep learning networks, stratified by two classes, basic services and value-added services. The most general "business model" in blockchain networks is a peer-based transaction fee for services or other credit assignation mechanisms, and in deep learning networks is a loss function optimization.

Basic services include administrative services that are expected for the orderly and secure operation of the distributed system. These services could include transaction confirmation, network security, governance, mining, and consensus algorithms, record-keeping (ledger-hosting), and wallet services (addresses, public–private key pair issuance). Another tier of services, value-added services, could run as a network overlay. Value-added services might include peer-based hosting of news and social

Table 12.6. Blockchain network services provided by peer-to-peer nodes.

Network services	Type of service provided by peer-to-peer nodes	Specific solution(s)
Basic services	Transaction execution, network security, wallet functionality (key pairs, addresses)	Transaction confirmation, network security
	Record-keeping, archival	Ledger hosting, updates
	Data storage	Storj, IPFS/Filecoin
	Software updates	Core developers, improvement proposals
	Routing (routing protocols arc open, just as internet routing protocols (TC/PIP))	Basic routing and smart routing services based on cost, speed, confidentiality
	Governance	Voting, participation, credit assignment, remuneration
Value-added services	News, social networking	Steemit, Yours
	Intellectual property registration	Proof of Existence, Monograph, Verisart, Crypto-Copyright
	Banking and finance	Payment channels (Lightning), credit (Ripple)
	Digital identity management, reputation services	Sovrin, Hyperledger Indy, Evernym, uPort
Token services	Local economy: community energy-sharing	Transactive Grid (Brooklyn NY), ION
Token services	Local economy: voting, community management	District0x, DAOdemocracy, Neon District, OpenSea.io
	Transportation	Lazooz
	Education certification (diplomas, transcripts)	MIT digital certs (diploma certificate)

networking applications, intellectual property registration (with hashing and time-date stamping), and digital identity management and reputation services. Banking and financial services could be provided via payment channels, and credit (open credit links on the Ripple network and in payment channels on the Lightning Network). Digital asset inventories

Table 12.7. Deep learning services provided by perceptron nodes.

Class of service	Type of service provided by perceptron nodes	Applications
Basic services	Data classification (train on existing data)	Facial recognition, language translation, natural language processing, speech-to-text, sentiment analysis, handwriting recognition
	Feature identification, multiple levels of abstraction	
	Data identification (correctly identify test data)	
Value-added services	Pattern recognition	Medical diagnostics, autonomous driving
	Optimization, error correction	Supply chain analytics
	Time-series forecasting, prediction, simulation	Error contribution
	Data automation	Privacy-protected data markets
	Memory functionality, sequential data processing	LSTM RNNs
Advanced	System architecture design and improvement	Dark knowledge
Advanced	Advanced learning techniques	Adversarial networks, reinforcement learning

(registration, tracking, pledging, and contracting) could be another service. Community participation and orchestration is another value-added network service. Services could include token issuance and management, custody, resource access, voting, and supply-and-demand matching. An example of local economy management services could be having an Uber-type app to coordinate the local peer-based economy for microgreen harvesting or other on-demand goods and services.

Deep learning perceptron networks could likewise have a number of basic services and value-added services. Basic services could include the general functionality of deep learning networks in object identification (IDtech) and data classification algorithms. Value-added services could include the next levels of deep learning functionality such as pattern recognition, optimization, and forecasting with LSTM RNNs. Advanced value-added services could include new methods such as adversarial networks, dark knowledge, manifold learning, and reinforcement learning, as well as self-criticality evaluated system architecture.

12.6.2 *Basic administrative services*

Generalizing from the blockchain and deep learning examples, a SNFT could be used to aggregate and manage a variety of administrative activities. This could be the basic class of administrative services that any smart network would be expected to have. Some of the most generic administrative activities include security, privacy, activity logging, and system maintenance. Within security, there could be identity confirmation, antivirus-type services, overall network security, and computational proofs. Within activity logging, there could be operational execution, secure audit-logging, backup, lookup and information retrieval, search, and data aggregation. Within system maintenance, there could be software updates, hardware scans, and other kinds of necessary system repair and maintenance.

12.6.3 *Value-added services*

At another level, there is a class of applications enabled by SNFTs that facilitates the practical engagement of the network to develop and provide more complicated value-added services. This level of activity enrichs and makes the network and its activities more substantive. The perceptron Hamiltonian for example, could signal factors of novel emergence that could then be translated into new value-added services or feedback to be incorporated into the deep learning network. On-demand services could fulfill needs ambiently on large or small scales (rolling trends in demand into network wide services).

12.6.3.1 *Systemic risk management*

An important class of value-added services is risk management. Systemic risk has been shown to increase as a function of the coupling strength between nodes (Battiston *et al.*, 2007). Hence, one suggested risk-management service would be using SNFTs to obtain a global measure of coupling strength between nodes, and understand how this might directionally impact systemic risk (Swan, 2018b). A problem that might be helped with SNFTs is financial instability in the form of flash crashes, which occur at faster-than-human-manageable time scales (Johnson *et al.*, 2013). Some of the proposed solutions for managing flash crashes include using statistical models for phase transition identification and Jones polynomials (which model the stock market with the technophysics model of a knot and braid topology) (Racorean, 2014). These methods could be applied together with field-theoretic principles to model and control smart network systems. The aim is to avoid local minima in an energy landscape, behavior such as double-pendulum chaoticity that could result in flash crashes (dos Santos, 2019). SNFTs could be used for the risk management of these kinds of excess effects of technology.

12.6.3.2 *Collective intelligence as an emergent system property*

Collective intelligence might be harnessed as an emergent system property, and provisioned as a resource across the network. Deep learning chains could be used to provide pattern recognition and privacy-protected computing services such that agents could have validation about the aggregated use of their information. One example could be a blockchain Hamiltonian for market sentiment. Another means of harnessing collective intelligence could be via the self-play method, as is used now between human agents, and in video games and adversarial deep networks. To reduce systemic risk, trade bots could be pitted against one another in a public goods game in which the (remunerated) goal would be avoiding system criticality, such as flash crashes. Human traders and investment houses could pay a small insurance fee per trade (commission) into the incentive pool for programmatic trading entities to prevent financial contagion. Collective intelligence could become a provisionable network resource for systemic risk management. The benefit of such a CrashWatch metric could be earlier warning signals in the case of market crashes, bringing more resolution to their effective management long before system criticality (a crash) happens. In another smart network system, UAVs, collective intelligence is being harnessed as an emergent system property.

A blockchain-based reinforcement learning model for air traffic control has been proposed based on these principles (Duong *et al.*, 2019).

12.6.3.3 *Network resource provisioning*

Another important class of value-added services is resource discovery and production. Field theories could help by detecting emergence in the network, not only for system criticality and risk management purposes, but also for new resource identification. SNFTs might be employed to find, measure, create, and deploy network resources. First, it is necessary to locate, measure, and deploy existing network resources. Second, it is necessary to identify emerging network resources and new layers (and types of layers) in the network stack. Third, for the generation of new resources, field-theoretic principles could be used to design, test, and simulate new resources before they are widely deployed to the network. For example, as smart network nodes (deep learning perceptrons and blockchain peer-to-peer nodes) might become more autonomous, the idea would be to canvas the nodes to gauge the network demand for deploying the new resources and previewing what new kinds of services might be delivered with the resources. Having more proactive and real-time resource-demand forecasting is one way that *supply chains* could shift to becoming *demand chains* instead. One such "demand chain application" could be predicting demand from user attributes and aspirations which are shared in privacy-protected Data Markets. These kinds of next-generation social networking applications have been articulated as value-added overlays to social network infrastructure (Swan, 2018a).

The smart network stack could have four primary layers: infrastructure, interface, application, and network resources. A smart network field-theoretic function such as a Boltzmann ensemble distribution (a probability distribution over all possible states of the system) could be used to measure the availability of network resources across layers. Just as synaptic excitation is a network resource that triggers the action (the motion of the overall network), so too are smart network resources such as algorithmic trust, information attributes, and liquidity (for example, automatic transaction financing and supply-demand matching). Another smart network field-theoretic function could be used to mobilize resources as the *propagation of certainty*. The idea is to take the inverse of the Gibbs measure of the *propagation of uncertainty* as a measurable quantity that is diffusing through the network. The *propagation of certainty* sums, measures, and calculates the quantity of available network resources, such as algorithmic trust. Certainty is a meta-resource, an information layer about the surety of the availability of resources, which targets the same need as a futures contract (guarantee of future resource availability) with a smart network market mechanism. Another new measure could be *platform entropy* as an information-theoretic measure of innovation. An increase in platform entropy (measured as an increased number of possible system configurations) is a metric indicating an increase in innovation.

12.6.4 *Smart network metrics*

12.6.4.1 *Economic indicators of the future*

The amount of available network resources (systemwide quantities) might be calculated as future Economic Indicators. The idea is to use SNFTs to measure network resource availability, and mange (or have self-managed) economic systems based on this. Ripple already constitutes a new form of real-time economic indicator, in the sense that it is a live credit network. A similar indication is available for the public channel activity on the Lightning Network. Open credit links create a live network for monetary transfer and serve as a real-time measure of economic confidence beyond any currently existing metrics. As credit-extending nodes retract their links, the measure of economic confidence and credit availability decreases (Swan, 2019). The credit elasticity of the Ripple network could be a smart network field-theoretic calculation that includes temporal aspects as live credit varies over time. Likewise, the idea of *real-time balance sheets* (as many high-value worldwide assets inventories might become blockchain-registered) could provide a stock market-like immediate valuation of assets, as opposed to calculations only performed after the end of the accounting period. Disclosure is to be privacy-protected. A first step could be shifting current economic indicators to real-time calculability with tools such as SNFTs. A basic smart network functionality is being able to measure and predict the amount of network resource availability in the past, present, and future. At minimum, the smart network should be able to signal information about resource availability and cost.

12.6.4.2 *Example: Maersk path integral for shipment networks*

Another example of a future Economic Indicator could be a real-time supply chain metric of the cost to ship a kilogram of goods per kilometer of distance worldwide. This could be called the "Maersk path integral," reflecting the dynamism and interrelation of potential shipping paths, and the possible quantum computational calculation of the optimized route. The Maersk path integral could be similar to the metric of the cost to lift a kilogram of cargo to space (Dreyer *et al.*, 2011). Over time, SpaceX has lowered this metric by an order of magnitude, citing a price of $4,700/kg for the existing Falcon 9 and $1,700/kg for future versions of the Falcon Heavy (Rockets, 2015).

Defining metrics can be a galvanizing step in focusing an industry's attention on improvement. An example of this is the cost to sequence a single human genome, which continues to motivate orders-of-magnitude improvement. Using SNFTs to define and measure metrics could be important for future-class projects. Some of the farther consequences of SNFTs could be using the theories as a blueprint, meaning as a formal means and planning mechanism, for defining future projects. As a conjecture, this could include employing SNFTs towards planning for the realization of higher-level scales of Kardashev societies (the ability of civilizations to create and deploy advanced technologies). The point is to have a comprehensive theoretical tool for very large-scale next-generation planning for advanced technologies. From a practical SNFTs perspective, the important short-term objective is to be able to characterize, monitor, and control smart network systems.

References

Andrychowicz, M., Denil. M., Gomez. S. *et al.* (2016). Learning to learn by gradient descent by gradient descent. arXiv:1606.04474 [cs.NE].

Battiston, S., Delli Gatti, D., Gallegati, M. *et al.* (2007). Credit chains and bankruptcy propagation in production networks. *J. Econ. Dyn. Control.* 31(6):2061–84.

Creswell, A., White, T., Dumoulin, V. *et al.* (2018). Generative adversarial networks: An overview. *IEEE Signal Processing*. 35(1):53–65.

dos Santos, R.P. (2019). Consensus Algorithms: A Matter of Complexity? In: Swan, M., Potts, J., Takagi, S., Witte, F. & Tasca, P. (eds). *Blockchain Economics*. London, UK: World Scientific.

Dreyer, L., Bjelde, B., Doud, D. & Lord, K. (2011). SpaceX: Continuing to drive launch costs down and launch opportunities up for the small sat community. In: *25th AIAA/USU Conference on Small Satellites*, pp. 1–6.

Duong, T., Todi, K.K. & Chaudhary, U. (2019). Decentralizing air traffic flow management with blockchain-based reinforcement learning. Aalto University, Finland.

Harrison, J. (1999). Flux across nonsmooth boundaries and fractal Gauss/Green/Stokes' theorems. *J. Phys. A: Mathematical and General*. 32(28):5317.

Hinton, G., Vinyals, O. & Dean, J. (2014). Distilling the knowledge in a neural network. In: *NIPS 2014 Deep Learning Workshop*. arXiv:1503.02531 [stat.ML].

Johnson, N., Zhao, G., Hunsader, E. *et al.* (2013). Abrupt rise of new machine ecology beyond human response time. *Sci. Rep*. 3:2627.

Kelly, F. (2008). The mathematics of traffic in networks. In: Gowers, T., Barrow-Green, J. & Leader, I. (eds). *The Princeton Companion to Mathematics*. Princeton University Press, pp. 862–70.

Miikkulainen, R., Liang, J., Meyerson, E. *et al.* (2017). Evolving deep neural networks. arXiv:1703.00548 [cs.NE].

Ni, S.X., Pan, J. & Poteshman, A.M. (2008). Volatility information trading in the option market. *JOF* LXIII(3):1059–91.

Racorean, O. (2014). Braided and knotted stocks in the stock market: Anticipating the flash crashes. arXiv:1404.6637 [q-fin.ST].

Rockets (2015). SpaceX: Reducing the cost of access to space. Harvard Business School.

Sole, R.V., Manrubia, S.C., Luque, B. *et al.* (1996). Phase transitions and complex systems. *Complexity* 13–26.

Swan, M. (2018a). Blockchain consumer apps: Next-generation social networks (aka strategic advice for Facebook). *CryptoInsider*.

Swan, M. (2018b). Blockchain economic networks: Economic network theory of systemic risk and blockchain technology. In: Treiblmaier, H. & Beck, R. (eds). *Implications of Blockchain*. London, UK: Palgrave Macmillan.

Swan, M. (2019). Blockchain Economic theory: Digital asset contracting reduces debt and risk. In: Swan, M., Potts, J., Takagi, S., Witte, F. & Tasca, P. (eds). *Blockchain Economics*. London, UK: World Scientific.

Part 5

The AdS/CFT Correspondence and Holographic Codes

Chapter 13

The AdS/CFT Correspondence

Abstract

The anti-de Sitter space/conformal field theory (AdS/CFT) correspondence is the theory of a proposed duality between a bulk region with (d+1) dimensions and a boundary region with (d) dimensions. The AdS/CFT correspondence (also called gauge/gravity duality) suggests that in any physical system, there is a correspondence between a volume of space and its boundary region such that the interior bulk can be described by a boundary theory in one fewer dimensions. The AdS/CFT correspondence is a formalization of the holographic principle which denotes the possibility of a 3D volume being reconstructed on a 2D surface. There is an information-theoretic interpretation of the AdS/CFT correspondence as a quantum error-correction code, which is formally solved with tensor network models. Many other fields use the AdS/CFT correspondence model to study various phenomena in superconducting materials, condensed matter plasmas, and network theory.

13.1 History and Summary of the AdS/CFT Correspondence

The key moments in the history of the anti-de Sitter space/conformal field theory (AdS/CFT) correspondence are encapsulated in Table 13.1. The ideas related to the holographic principle and the AdS/CFT correspondence began with Bekenstein and Hawking wondering about how

Table 13.1. Key historical moments in the AdS/CFT correspondence.

Bekenstein-Hawking entropy formula (1973, 1975)	Black hole entropy scales by area not volume
Holographic principle (Susskind ('t Hooft), 1995)	Complementary views of the same physical phenomena
AdS/CFT correspondence (Maldacena, 1998)	Bulk/boundary correspondence (gauge/gravity duality)
AdS/CFT entanglement entropy formula (Ryu & Takayanagi, 2006)	Boundary entanglement entropy related to bulk minimal surface
MERA tensor networks for quantum mechanics (Vidal, 2008)	Tensor network formulation for quantum mechanical entanglement
Apply MERA to Ads/CFT (Swingle, 2012)	Model the AdS/CFT correspondence as an entangled quantum system
AMPS thought experiment: Black hole firewall paradox posed (Almheiri *et al.*, 2013)	Claim: information is knowable about outwardly radiating bits from a black hole
The AdS/CFT correspondence is an information theory problem (Harlow & Hayden, 2013)	Claim: even with a quantum computer, cannot determine information about outwardly radiating bits from a black hole
Interpretation of AdS/CFT as a quantum error-correcting code (Almheiri *et al.*, 2015)	Quantum error correction as a model for the AdS/CFT correspondence
Exact solution of AdS/CFT as quantum error-correcting code (Pastawski *et al.*, 2015)	Formalizing a specific holographic quantum error-correction code

information is treated in black holes, and posing the black hole information paradox. On the one hand, black holes are a singularity at which known laws of physics break down, but on the other hand, maybe not all principles of physics are invalidated, such as thermodynamics. Moreover, since black holes are known objects in the universe, a question arises about the relationship between black holes and non-black hole regions. These questions led to the finding that black hole entropy is different from regular entropy. Black hole entropy scales by area as opposed to volume, whereas thermodynamic or von Neumann entropy scales by volume. For example, if a room were to be filled with computer hard drives, the amount of information that could be stored is based on the volume of the room, not the area, but in black holes, the entropy is related to the area.

The so-called entanglement area law refers to entropy that scales by area not volume (Bekenstein, 1973).

The next important finding is that the entanglement area scaling law applies to any quantum mechanical domain more generally, including quantum many-body problems, quantum information science, and black hole physics (Eisert *et al.*, 2010). In quantum systems, entropy scales by area not volume. The important practical implication is that quantum mechanical problems immediately become much easier to compute, since area is easier to calculate than volume. Scaling by area not volume means that previously intractable numerical analysis models of quantum systems are in fact possible, particularly with tools such as tensor networks. A crucial advance is the Ryu–Takayanagi entanglement entropy formula that implements the Bekenstein–Hawking area entropy in the Ads/CFT correspondence (Ryu & Takayanagi, 2006). In the Ryu-Takayanagi formula, entanglement entropy in the boundary is computed by the area of a certain minimal surface in the bulk geometry. Bulk entropy is black hole or quantum mechanical-type entropy that scales by area not volume, whereas boundary entropy is "traditional" thermodynamic or von Neumann-type entropy that scales by volume. The difference in bulk entropy and boundary entropy alone implies many interesting problem-solving applications.

One puzzle about the fact that bulk entropy and boundary entropy are different relates to how information moves between the two domains. New forms of the black hole information paradox point out that information is radiated out of black holes, and ask how this could be possible since nothing can escape from a black hole. Perhaps there are two versions of the information, one that goes into the black hole and one that radiates out, but this would violate the no-cloning theorem of quantum information (quantum information cannot be copied). As a solution to this, Susskind (extending the ideas of 't Hooft) proposed the holographic principle (Susskind, 1995). The holographic principle is that there are different but complementary views of the same physical phenomenon. There are not two instances of the information, but two different viewpoints from which the same information can be seen. From relativity, it is known that a phenomenon looks different to different observers. A far-off observer only sees information smearing out on the event horizon of the black hole (the boundary) and never actually entering the black hole, whereas a near-by observer that is jumping into the black hole sees the information going into the bulk. Each could analyze the entropy per their own stance, with the area entropy law in the bulk, and the volume entropy law on the boundary. Due to the complementary views of observers per the holographic principle (far-off in 2D and near-by in 3D), there is no paradox.

Maldacena relates the bulk and boundary ideas to establish a formal connection between the two domains, and defines the AdS/CFT correspondence (Maldacena, 1998), which might be more accurately termed gauge/gravity duality (Maldacena, 2012). Various theorists employ the AdS/CFT correspondence in different ways. Harlow & Hayden (2013) propose information theory and computational complexity as a way of thinking about the black hole firewall paradox (the "AMPS" thought experiment) posed by Almheiri *et al.* (2013). The black hole firewall paradox is an extension of Hawking's black hole information paradox (1975), which asks how entangled bits of quantum information can radiate out of a black hole. Harlow & Hayden claim that even with a quantum computer, there would not be enough time to calculate information related to the entanglement entropy of an information bit radiating out of a black hole (countering the AMPS claim that this would be possible). The advance of Harlow & Hayden is suggesting a link between physics and information theory, and using computational complexity to analyze these problems. Quantum error correction, zero-knowledge proofs (in the form of Quantum Statistical Zero-Knowledge, QSKP), and quantum secret-sharing are elements of quantum information theory that have a bearing on the problem.

Almheiri *et al.* (2015) extend the ideas related to information theory and computational complexity by interpreting the AdS/CFT correspondence as a quantum error-correcting code. Then, Pastawski *et al.* (2015) offer a realization of Almheiri *et al.*'s proposal by implementing the AdS/CFT correspondence formally with tensor networks. Pastawski *et al.* propose the specific structure of a holographic code that can be used for error correction between bulk and boundary regions. The AdS/CFT correspondence is modeled with a MERA-like tensor network (MERA: multi-scale entanglement renormalization ansatz) as proposed by Swingle (2012). Tensor networks are a computational tool used for instantiating quantum many-body problems, and MERA specifically incorporates the entanglement property of quantum systems (Vidal, 2008).

Research continues on the issues raised in this body of work. Harlow proposes an algebraic quantum error-correction method for the Ryu–Takayanagi formula for entanglement entropy (Harlow, 2017). In other research, Almheiri suggests a holographic quantum error-correction mechanism for the black hole interior (Almheiri, 2018). Dong *et al.* define a de Sitter space interpretation (rather than an anti-de Sitter space interpretation) of the AdS/CFT correspondence (Dong *et al.*, 2018). This could be useful as the regular space of lived existence is de Sitter space, whereas anti-de Sitter space is a specialized experimental formulation. Other work by Osborne & Stiegemann (2017) introduces dynamics for the holographic codes as proposed by Pastawski *et al.* (2015). Some of the various tools and methods used in this line of AdS/CFT correspondence research are listed in Table 13.2.

Table 13.2. Tools and methods used in AdS/CFT correspondence research.

Harlow & Hayden (2013)
• Holographic principle, AdS/CFT correspondence
• Information theory
• Computational complexity classes (QSZK)
• Quantum error correction
• Thought experiments
Almheiri *et al.* (2013)
• Holographic principle, AdS/CFT correspondence
• Quantum error correction
• Spatial transformations (Ryu–Takayanagi entanglement entropy, radial transforms, and Bogoliubov transformations)
• Bulk reconstruction via causal wedge (AdS–Rindler reconstruction) and entanglement wedge
• Cauchy surfaces
• Operator algebra
• Quantum secret-sharing
Pastawski *et al.* (2015)
• MERA-like tensor networks
• Quantum error correction
• Holographic codes

13.2 The AdS/CFT Correspondence: Basic Concepts

13.2.1 *The holographic principle*

The holographic principle was proposed by Susskind, building on work from 't Hooft (Susskind, 1995). The holographic principle is the general phenomenon of a physical theory that can be written in some number of dimensions (e.g. the bulk volume of a 3D space), which turns out to be dual in some sense to a completely different physical theory defined on the boundary of that region (a 2D surface). The two theories are different but comprise two ways of looking at the same physical situation. The duality between them is that there is a one-to-one mapping between the states of the first theory and the states of the second theory. For example, there could be a particle inside of the bulk, which in the boundary theory might correspond to an enormous smeared out object. This is how quantum error correction is performed, by smearing out the information to be protected (the logical qubit) across extra qubits (physical qubits).

The holographic principle refers to the notion of a hologram, in which a 3D image is recorded on a 2D surface. Lasers are used to create interference patterns. A hologram captures the interference pattern between the laser beams and records it on the recording medium, which is a 2D surface. Later, when the hologram is lit up in a certain way, the recorded pattern can be seen by an observer.

In a hologram, the idea is that a 3D image is stored on a 2D space, or more generally that a 2D surface can represent a 3D volume. A (d) dimensional surface can represent a (d+1) dimensional volume; the boundary is the representation of the bulk in one fewer dimensions. The key point is that the information needed to describe a 3D volume can be compressed into a 2D surface, and the compressed information on the 2D surface can be used to reconstruct the information in the 3D volume. One can imagine a bug being smeared out on a car's windshield. The bug had a 3D existence, but is now stored on a 2D plane. All of the information is still there, it is just smeared out on a plane in one fewer dimensions.

Reconstruction is an important concept in the AdS/CFT correspondence, in that one region can be used to reconstruct the other. Either the

3D volume can be reconstructed on the 2D surface (in one fewer dimensions), or the 2D surface can be reconstructed (interpreted) in the 3D bulk (as a minimal surface).

13.2.2 *Holographic principle formalized in the AdS/CFT correspondence*

The holographic principle is made more precise in the AdS/CFT correspondence (Maldacena, 1998, 2012). As an instantiation of the holographic principle, the AdS/CFT correspondence is a duality between a bulk with (d+1) dimensions and a boundary with (d) dimensions. One theory describes the bulk or the interior, and is an effective field theory that includes gravity (in the general formulation, the bulk effective field theory is a quantum theory of gravity). Another theory describes the boundary, and is an ordinary conformal field theory (meaning having no gravity and a flat space–time). Conformal field theories are specific examples of field theories, in one fewer dimensions. A conformal field theory is a robust general formulation for the boundary because it is a quantum field theory that is invariant under certain kinds of changes. More technically, a conformal field theory is invariant to conformal transformations.

Conformal transformations are mappings that preserve local angles (leaving the size of the angle between corresponding curves unchanged). Conformal field theories are used in the AdS/CFT correspondence to articulate that a gravitational theory in the AdS bulk is equivalent to a conformal field theory in the boundary. The point is that the two regions are related in the AdS/CFT correspondence, the bulk interior of the volume and its flat boundary region on the edge. They have a relationship (a duality) that can be formalized. The bulk and the boundary are not just randomly connected, but there is a certain mapping that can be made between the two regions or theories. The holographic correspondence is most specifically conceived as the mapping between conformal field theories and higher dimensional theories of gravity.

Sometimes the AdS/CFT correspondence is analogized to a soup can in that the bulk is the 3D interior of the can and the boundary is the

2D exterior of the can. This is a useful visual representation of the difference between a 2D surface and a 3D bulk, however an important point is that AdS is referring to anti-de Sitter space as opposed to regular de Sitter space. Regular de Sitter space is the normal 3D space of lived reality, which can be described by Euclidean geometry. Anti-de Sitter space is based on the hyperbolic geometry of a sphere, and looks like the Circle Limits works of Escher that have pictures of fish and bats getting smaller and smaller as they extend towards the edge of a circle. Anti-de Sitter space is used in physics because it is a simpler toy model of de Sitter space. The formulations were proposed by de Sitter, a contemporary of Einstein's (de Sitter, 1917). AdS provides exact solutions to Einstein's field equations, but for an empty universe with constant negative scalar curvature (like the Escher drawings), which is not the standard model of the universe which has a positive vacuum energy. However, the AdS/CFT correspondence is one of the best-understood models of a theory of quantum gravity even though it is a toy model of reality. The AdS/CFT correspondence is used in this context to study many bulk/boundary problems such as the dynamics of quantum field theories and the relationship between domains of weakly coupled gravity in bulk regions and strongly coupled fields in boundary regions.

13.2.3 *Quantum error-correction code interpretation*

The mapping between the bulk and the boundary regions can be expressed in different ways, through a correspondence of theories, states, operators, spatial structure, parameters (called degrees of freedom), or other elements. In terms of spatial structure, both regions are Hilbert spaces, but the boundary is a Hilbert space in one fewer dimensions. The mapping may be formalized in a dictionary that translates the terms between the two regions such that they can be used for computation. For example, in quantum error correction, a dictionary mapping links the bulk logical qubit with the boundary physical qubits to perform quantum error correction. The dictionary mapping is a quantum error-correcting code because it expresses a code (a mechanism) for performing the error correction between the two regions.

13.3 The AdS/CFT Correspondence is Information-Theoretic

A key advance of Harlow & Hayden (2013) is seeing the AdS/CFT correspondence as an information-theoretic problem. Black holes can be conceived as a quantum mechanical domain with entangled qubits that can be error-corrected. The problem is as follows. Over time, black holes evaporate, by radiating the qubits of information that comprise them back out into the universe. This is called Hawking radiation. The claim is that information is not lost, even in a black hole, although it may be difficult to access. An analogy is made to a book that is burned in a fire. The information is still there, but it is very difficult to access. At any moment in time, all of the qubits related to a particular black hole will be in one of three locations, in the black hole interior, on the event horizon, or in the exterior of the black hole.

A question arises as to how quickly information can be obtained from a qubit that is entering or exiting a black hole. One way would be to have a giant tracking system that catalogs all of the qubits related to the black hole. Qubits have the property of being monogamously entangled with other qubits. Therefore, such a tracking system could assess whether an outgoing qubit is part of a pair whose first qubit has already been radiated out of the black hole (comprising a 2 of 2 pair), or is part of an as-yet unknown pair (comprising a 1 of 2 pair). Whereas the AMPS thought experiment (Almheiri *et al.*, 2013) suggests that these kinds of calculations would be possible, Harlow & Hayden (2013) apply an information-theoretic approach and conclude that the black hole would evaporate before such calculations would be possible, even with a quantum computer. More formally, the situation comprises the well-defined computational problem of a polynomial-sized quantum circuit acting on three systems (black hole interior, horizon, and exterior) to determine if maximal entanglement can be decoded, and it cannot (Harlow & Hayden, 2013, p. 7).

13.3.1 *Black hole information paradox*

Harlow & Hayden develop their ideas in the context of the black hole information paradox. The black hole information paradox is a modernization of Hawking's initial black hole information paradox, which asks how it is possible for information to exit from a black hole (1975). The conundrum results from trying to understand the combination of quantum mechanics and general relativity together. The paradox is that on the one hand, information apparently cannot escape from black holes (per general relativity), but on the other hand, Hawking radiation does emanate from black holes, and the only way the information can be in the Hawking radiation is if what is inside the black hole is copied, however, the no-cloning theorem prevents there from being two copies of the quantum information (per quantum mechanics).

The holographic principle (also called black hole complementarity) was proposed as a solution by Susskind, per the work of 't Hooft and others (1992). Black hole complementarity suggests that the information is both inside and outside the black hole, in a way that does not violate quantum theory (because it does not require two copies of the information). Specifically, far-off observers who remain outside of the black hole see the information accumulating or smearing out on the horizon, never actually entering the black hole, and then later, evaporating out in the Hawking radiation. Near-by observers who fall into the black hole see the information located inside the black hole. Since the two observers cannot communicate, there is no paradox. Hawking indicated agreement with the proposal (Hawking, 2005), supporting the idea that there is no information loss in black holes.

The complementarity proposal is not unreasonable given what is known of time dilation and special relativity. Any event, including black hole information evaporation, looks different to different observers in the universe. The far-off observer sees everything smeared out on the black hole event horizon, and the person falling into the black hole sees objects falling into the black hole. One can consider the situation of the near-by observer falling into the black hole with a book (i.e. information). From the point of view of the far-off observer, the information never goes beyond the event horizon, it is just pancaked out (smeared out) on the 2D event horizon, like a bug hitting a windshield. Due to time dilation, the far-off observer never actually sees the object fall into the black hole. The book gets slower and slower through time dilation as it nears the black hole, and more and more smeared out on the event horizon. From the point of view of the far-off observer, the interior of the black hole could be treated as not even existing (the event horizon is essentially a firewall which no information gets through). The point of view of the near-by observer is different, this observer has the 3D experience of falling into the black hole. This is how the holographic principle (two separate points of view) solves the apparent black hole information paradox. There are two different ways of looking at the same physical situation, the interior (bulk) 3D point of view and the exterior (boundary) 2D point of view in which everything is taking place on the surface of the event horizon. The same holographic principle applies whether information is entering or exiting the black hole.

13.3.2 *The information-theoretic view*

Since information evaporates out of the black hole in the form of Hawking radiation, and given the information-theoretic view that the book (i.e. black hole information) looks different to the far-off and the near-by observers, Harlow & Hayden posit an information-theoretic interpretation of the AdS/CFT correspondence. Such an information-theoretic view implies computational complexity, in that the computational resources (in time and space) required to calculate a given problem can be analyzed. Various classes of computational complexity apply to quantum information (such as the Hawking radiation exiting the black hole), including QSZK.

The question is how quickly information can be obtained from a qubit that is radiating out of a black hole. Casting this problem in the computational domain, the question is how quickly information can be computed about a qubit that is radiating out of a black hole, since the qubit cannot be measured directly without destroying it (per the properties of quantum information). One method for computing such information is by applying a quantum error-correction scheme.

Quantum error correction denotes the situation that since a qubit is entangled with another qubit, if the original qubit becomes damaged, it can be error-corrected (re-established in its original form) by performing a quantum error correction process with the entangled qubit. Quantum information cannot be measured directly (without destroying it), but information about 1 qubit may be recovered from another qubit with which it is entangled. Even though the outwardly radiating qubit is not destroyed and in need of error correction, the point is that the error correction mechanism allows information about the qubit to be obtained without measuring it directly. Quantum information problems imply entangled qubits and quantum error correction as a means of obtaining information about entangled qubits. However, one problem is that this particular error correction scheme will only work if the qubit with which the outwardly radiating qubit is entangled is already out in the universe somewhere and not still in the black hole interior.

These kinds of quantum error correction methods imply extraordinarily complex calculations, calculations that are in a complexity class such that the problem cannot be evaluated in a reasonable amount of time, even with a quantum computer. Harlow & Hayden show that because the error-corrected information is in the QSZK computational complexity class, it will not be possible to calculate the problem (perform the error correction) in polynomial time (a reasonable amount of time) on a quantum computer. In fact, the black hole will evaporate before the computation is complete.

The problem of obtaining information about an outwardly radiating qubit from a black hole is at least QSZK-hard (has the difficulty of calculating problems in the class of QSZK). Problems that are QSZK-hard cannot be solved in polynomial time on a quantum computer. QSZK (Watrous, 2002) is a standard computational class of quantum information that is basically a more complicated version of BQP, the class of all problems that can be solved on a quantum computer (Bernstein & Vazirani, 1997).

Recognizing that the problem is a form of QSZK helps to quickly determine that it cannot be solved in an amount of reasonable time (due to its complexity class). The benefit of the information-theoretic approach is that it provides a quantitative method for analyzing these kinds of physics problems. The method is a basis for extending problem-solving methods in this domain, and defeats the previous result in the research trajectory. The previous result theorizes that it would be possible to obtain information about an outwardly radiating qubit (in the so-called AMPS experiment, after the names of the authors of the paper (Almheiri *et al.*, 2013),

correspondence, and the code subspace is a further subspace within the AdS–Rindler wedge, in the bulk. Second, with the code subspace delineated, operators (logical operators) are defined to act directly on the code subspace to perform the error correction. The error correction is the usual formulation, a logical qubit in the bulk to be error-corrected with entangled physical qubits in the boundary. The bulk logical operators correspond to physical operators in the boundary. Third, a precision method for quantum error correction, operator algebra quantum error correction, is applied. The algebra can be restricted to a certain subalgebra of operators on the code subspace within the subregion–subregion correspondence (Beny *et al.*, 2007). The insight for quantum error correction is the idea that the radial (inward–outward) direction in the bulk is realized in the boundary as a measure of how well the boundary representations of bulk quantum information are protected from local erasures.

The result is that the bulk–boundary correspondence is defined in a new and refined way. The local operators in the bulk are established as being dual to operators acting in certain subspaces of states in the boundary. The error correction method allows this duality to be articulated, by defining a code subspace in the bulk, and logical operators that act on the codespace to error correct entangled regions in the boundary through corresponding boundary operators (generated in the quantum error correction operator subalgebra). Ultimately, there is a duality between the bulk and boundary operators in that the local bulk operators are dual in the boundary to the subalgebra quantum error correction operators. The error correction protocol could be implemented in different ways, for example as the recovery of states, the action of operators, or the reconstruction of structure.

13.4.3 *Emergent bulk structure through error correction*

The consequences of the information-theoretic interpretation of the AdS/CFT correspondence as a quantum error correction code are two-fold. First, the immediate problem is solved, in producing a more granular model in which only certain bulk and boundary operators commute, overcoming the implication that all operators would commute in the AdS/CFT correspondence. The more refined formulation is that only certain operators commute because only certain bulk subregions are defined that correspond with certain entangled subregions in the boundary. The benefit of quantum error correction as a model is that it provides a general perspective on the issue of bulk reconstruction and a specific framework for articulating and understanding subregion–subregion duality between bulk and boundary regions.

Second and more importantly is being able to use the more granularly defined correspondence as a laboratory for other problems. The error correction method defines the correspondence in analytic specificity and this can be extended to other situations, including, notably to understand more about the bulk. The central finding of this research is the claim that bulk locality is a statement about certain subspaces of states in the boundary, and that using the error correction method for defining the correspondence helps to demonstrate the idea that bulk geometry emerges from boundary entanglement. The error correction method more firmly links the bulk and the boundary regions, and because the boundary is linked to the bulk, the correspondence can be used to learn new things about the bulk. This is a key target of physics research, understanding the interior of quantum many-body systems ranging from black holes to quantum computing environments.

13.4.3.1 *Implications of the correspondence as error correction*

The bulk–boundary correspondence might be applied to various situations. The correspondence was introduced to consider black holes, articulating a holographic correspondence between the 3D interior bulk and the 2D surface of the event horizon. The overall universe also might be examined with the correspondence, in the idea that perhaps the universe comprises the bulk, and has some surface on the edge. Even without being able to access the edge, the correspondence hypothesis is that bulk structure can be explained in one fewer dimensions in the edge. The correspondence might likewise be used in quantum many-body systems, such as those in quantum computing, as a tool for rapid portability in dimension-spanning to analytically simplify and understand how such complex systems are functioning.

One research question for understanding a more complicated system from a less complicated boundary's-eye view is framed as understanding emergent locality in the bulk, how local structure emerges in a bulk region. The insight from the holographic principle is that it may be easier to understand a phenomenon in one fewer dimensions. The hypothesis is that the bulk structure can be understood in one fewer dimensions in the way it appears on the boundary (in the corresponding boundary structure). Further, having or obtaining boundary structure in one fewer dimensions suggests being able to reconstruct the bulk structure in one greater dimension (the 3D bulk can be reconstructed from the 2D surface). The potential solidity of this method is due to the correspondence, which suggests formally writing the terms of one region in the corresponding terms of the other. One of the most important problems is understanding how geometric structure (geometry) arises in the bulk, because geometry gives rise to space and time, according to some theories (space and time are perhaps not fundamental but emergent). For some theorists, the central research problem in the Ads/CFT correspondence is understanding the emergence of bulk locality, specifically bulk geometry (and space and time) from boundary entanglement.

Whereas existing research has focused on spatial transformations between the bulk and the boundary (Bogoliubov transformations, radial directionality, and Ryu–Takayanagi bulk–boundary entanglement entropy (Almheiri *et al.*, 2015)), temporal transformations are likewise an opportunity in AdS/CFT correspondence research. The AdS/CFT correspondence is implicated in articulating the emergence of bulk geometrical structure which includes both space and time. The global reconstruction that instantiates the bulk/boundary correspondence gives the boundary a time dimension (time complexity) by defining the boundary as a Cauchy surface (a plane with a time dimension). Hence, the time dimension can also be manipulated in the AdS/CFT correspondence. Analytic tools such as MERA tensor networks and random tensors (designed to connect regions with geometry and regions with subatomic particle interaction) might support the practical exploration of such correspondence in temporal transformations. One proposal along these lines is the idea of constructing 1D conformal field theories from tensors that only depend on the time variable (Witten, 2016).

13.4.4 *Extending AdS–Rindler with quantum secret-sharing*

The notion of the AdS–Rindler reconstruction as a model for quantum error correction might be applied in different ways. One idea is that the reconstruction could be extended if the quantum error correction can be instantiated as a quantum secret-sharing scheme.

Quantum secret sharing is a standard method used in cryptography (Cleve *et al.*, 1999). The notion is to have a threshold scheme in which a secret quantum state is divided into n-shares such that any k of those shares can be used to reconstruct the secret, but any set of k–1 or fewer shares cannot. The core concept is essentially the same as a quantum error correction scheme, in which a smaller set of information (a logical qubit) is recapitulated from a sufficiently larger set of information (entangled physical qubits).

Although similar in concept, quantum secret-sharing is more secure, and expands the potential reach of regular quantum error correction. The premise of a quantum secret-sharing scheme is that a larger domain of errors can be efficiently corrected if the partial information needed for the error correction can be shared across more parties. The key point of the scheme is the threshold property that a certain threshold must be reached to perform the action. The multi-party threshold property has many uses in encryption schemes (for example in digital signatures in addition to error correction).

The AdS–Rindler reconstruction might be extended in scope by instantiating the quantum error correction as a quantum secret-sharing scheme. The reach of the error correction (and also the correspondence between bulk and boundary regions) might be broadened by introducing a quantum secret-sharing requirement in the error correction. A wider range of regions could perform error correction indirectly via a quantum secret-sharing scheme than through a direct quantum error correction scheme. Whereas the Rindler causal wedge defines a relatively small zone of direct causal relationship between bulk and boundary regions, an alternative method, an entanglement wedge, might be defined for an even larger zone of bulk–boundary correspondence that can be related through a quantum secret-sharing scheme. The upshot is that in the context of the AdS/CFT correspondence, quantum secret-sharing through thresholding is an analytic tool that might be employed to realize a wider range of precision correspondence between the bulk and the boundary regions.

13.5 Holographic Methods: The AdS/CFT Correspondence

Holographic methods and the AdS/CFT correspondence specifically, are emerging as a general approach with wide-ranging use. The AdS/CFT correspondence (also called gauge/gravity duality) has been applied to many fields including basic physics research, superconducting materials, condensed matter, strongly coupled plasmas, and anomaly detection (Natsuume, 2015). The correspondence is proving to be a standard model for interrogating complicated bulk domains (any kind of spatial volume) with a surface theory in one fewer dimensions. One area of research is ongoing work on the correspondence itself. Another focus is in astrophysics and cosmology, since the holographic correspondence arose as a purported mapping between quantum field theories and higher dimensional theories of gravity. One area of study is previously unknown forms of macroscopically entangled quantum matter found in stars, black holes, and other cosmological systems.

A key advance of the correspondence is a change in overall perspective. Previously, scientists scratched their heads about the possibility of trying to establish a grand unified theory that links general relativity and quantum mechanics, or disclaimed that unification was a useful project. The reframing of the perspective is that aspects of particle theories (quantum mechanics) and gravity (general relativity) are jointly implicated in many phenomena, particularly in condensed matter physics, and hence focusing on these specific problems instead of a larger overall attempt at unification. The term gauge/gravity duality is indicative of the shift in viewpoint in that gauge theories refer to the behavior of subatomic particles, for whom gravity (geometric aspects) is relevant, in condensed matter for example. In the practical use of the correspondence, gravity is either meant literally, or as geometry, connoting geometric aspects, and space and time.

An example of the shift in perspective to gauge/gravity duality that involves both particle physics and geometric aspects can be seen in the study of condensed matter physics. Processes for analyzing condensed matter physics often involve using a model of finite charge density in order to stabilize the system (Ammon & Erdmenger, 2015). This applies to analyzing both so-called Fermi surfaces and condensation processes. In the gauge/gravity duality context, a model for finite charge density can be obtained by considering certain kinds of charged black holes (e.g. a frequently studied type of black holes called Reissner–Nordstrom black holes). The gravity action of these black holes involves additional gauge fields (subatomic fields), which is a model that can be used in solving condensed matter problems. The gauge/gravity duality approach is used to calculate standard thermodynamical quantities in condensed matter such as free energy and entropy, and more interesting parameters such as frequency-dependent conductivity.

13.5.1 *The correspondence as a complexity technology*

The correspondence is a general tool for framing problems as those consisting of a more complicated bulk domain, which can be interrogated in one fewer dimensions on the boundary, and also as a complexity technology (ComplexityTech). The correspondence can be seen as a complexity technology, meaning a tool that can be used to study and manage aspects of complex systems such as anomaly, critical points, phase transitions, and symmetry-breaking. Complexity is a feature of quantum systems. The emergence of many-body effects can give rise to symmetry-breaking phase transitions. These phase transitions could be ordered states (magnetic, charge, or superconducting) produced as the result of electron correlations (when each electron's behavior in its surroundings is influenced by the presence of other electrons).

The AdS/CFT correspondence is engaged to study complexity in quantum systems. One example is using the correspondence to investigate conformal anomalies (scale anomalies that lead to symmetry-breaking) (Ogushi, 2001). Another example is using magnetic fields and vortices as a control parameter to study quantum phase transition within gauge/gravity duality (Strydom, 2014). Standardized treatments are emerging such as renormalization, for example by providing a single computational framework for the renormalization group in the context of gauge/gravity duality

(Zaanen *et al.*, 2016). Since traditional methods may fail in the study of complex quantum systems, other (nonlinear) methods are indicated. Perturbation is one such approach (Fruchtman *et al.*, 2016) and holographic models is another.

13.5.2 *Strongly coupled systems: AdS/CMT correspondence*

One of the biggest topics on which the AdS/CFT correspondence is being directed is to study strongly coupled systems. Strongly coupled systems are quantum mechanical systems with strong interactions between particles, as determined by one of the four fundamental forces. Strongly coupled systems are not well understood, but are implicated in many phenomena, including in the superconducting materials used in quantum computing. The gauge/gravity duality is being applied to strongly coupled systems in the areas of condensed matter physics and plasmas. A subfield of study with the AdS/CFT correspondence has arisen, anti-de Sitter/condensed matter theory (AdS/CMT) correspondence to study holographic quantum matter (Pires, 2014; Hartnoll *et al.*, 2018).

13.5.2.1 *Coupling in physical systems*

In physics, two objects are said to be coupled when they are interacting with one another. In classical mechanics, an example of coupling is the relationship between two oscillating systems, such as pendulums connected by a string. In quantum mechanics, two particles that interact with one another are coupled. The interaction is caused by one of the fundamental forces, whose strengths are usually given by a coupling constant. Forces that have a coupling constant greater than 1 are said to be strongly coupled, and those with constants less than 1 are said to be weakly coupled. The basic idea is that strongly coupled particles are experiencing a significant amount of fundamental force.

Strongly coupled systems present an analytical challenge in that the interactions between particles become so intense that traditional methods such as mathematical tools and perturbative treatment break down (reach singularity). Hence new tools are needed and starting to be employed such as holographic methods (Bigazzi *et al.*, 2016).

13.5.2.2 *Holographic methods*

The correspondence is being applied to the study of various problems in the area of strongly coupled systems and condensed matter physics. The general principle in applying the correspondence is to reduce the quantities to be computed to a situation of solving the more complicated equations (such as those related to classical gravitation) in one higher dimension (one fewer dimensions) than the original theory. The correspondence is useful because it provides a geometrical framework for examining strongly coupled field theories (Pires, 2014). One benefit of the correspondence is that it can incorporate more granular detail and complexity such that the most basic toy-model systems are no longer required to examine the complex systems and exotic phenomena comprised by strongly coupled systems. For example, more complex models with non-zero temperature and non-zero density are easily managed with holographic methods. The correspondence can be used to study finite temperature real-time processes, such as response functions and dynamics far from equilibrium in quantum critical points in condensed matter systems (Lopez-Arcos *et al.*, 2016).

13.5.2.3 *Holographic methods and superconducting*

The most immediate practical use of an improved understanding of strongly coupled systems and condensed matter physics is in relation to superconducting materials that are used in quantum computing. Holographic methods provide a more granular model that can be used to understand strange phenomenon in strange materials. Fermi systems, meaning Fermi liquids, gases, and surfaces are in general, exotic phenomena. Fermi liquids are a theoretical model of interacting fermions that describes the normal state of most metals at sufficiently low temperatures (i.e. superconducting temperatures). Hence, holographic descriptions of the thermal properties of matter can be used to study Fermi liquids in the situation of superconducting. An example is examining cuprates, a copper material used in superconducting, which has a strange metal region with anionic complexes (complexes of ions with net negative charges (having more electrons than protons)). Holographic methods might also be applied to study the thermo-electric transport mechanism (particularly momentum dissipation) in strongly coupled condensed matter systems such as high-temperature superconductors (Amoretti *et al.*, 2017). The research finds discrepancies between the experimentally measured transport properties of these materials and the predictions of the weakly coupled theory of Fermi liquid that is usually used to model them.

A Fermi surface is essentially the partition line between system phases based on the states of electrons, which is also implicated in superconducting. The shape of the Fermi surface is derived from the periodicity and the symmetry of the crystalline lattice and the specific occupation of electronic energy bands. Simple AdS duals may lead to improved theories of Fermi surfaces that relate to Ising phase transitions in metals (Sachdev, 2011). Another system of interest is an ultra-cold Fermi gas (as opposed to an ideal Fermi gas) and how such a gas becomes tuned to saturate unitarity scattering limits.

13.5.3 *Strongly coupled plasmas*

A plasma is one of the four fundamental states of matter (which are solid, liquid, gas, and plasma). Specifically, a plasma is an ionized gas consisting of positive ions and free electrons in proportions that generally result in no overall electric charge. There are two kinds of plasmas, weakly coupled plasmas and strongly coupled plasmas, depending on the strength of the electrical force of each atom that holds the plasma together. Weakly coupled plasmas appear at low pressures (as in the solar corona, the upper atmosphere, and in fluorescent lamps). Strongly coupled plasmas arise at very high temperatures (as in white dwarf stars, giant planets, and nuclear fusion reactors).

Strongly coupled plasmas are an extreme phenomenon of charged many-particle systems comprised of multiple components of electrons, ions, atoms, and molecules. Strongly coupled plasmas (such as quark-gluon plasmas) are produced and studied through heavy ion collisions at facilities such as the LHC. Strongly coupled plasmas present a research puzzle in that whereas ideal plasmas (the ideal formulation, like an ideal gas) have higher kinetic energy than potential energy, strongly coupled plasmas have higher potential energy than kinetic energy. Since detailed information about the physical properties (such as temperature and density) of strongly coupled plasmas is not available (the region constitutes a bulk volume), holographic methods are a viable option for modeling these systems.

Holographic methods have been applied to study strongly correlated plasmas. One project finds that in the case of strongly coupled condensed matter systems such in high-temperature superconductors, as compared with other methods, the holographic correspondence suggests a simplification in that just four phenomenological entries are needed to fully determine the six independent thermo-electric transport coefficients in a strongly correlated plasma (Amoretti *et al.*, 2017). Other research implements the AdS/CFT correspondence to compute the dynamical relaxation times for the chiral transport phenomena and spin polarization in strongly coupled plasma regimes (Li & Yee, 2018). The result is that the relaxation times can be a useful proxy for the dynamical time scale for achieving equilibrium spin-polarization of quasi-particles in the presence of magnetic fields and fluid vorticity.

References

Almheiri, A. (2018). Holographic quantum error correction and the projected black hole interior. arXiv:1810.02055 [hep-th].

Almheiri, A., Dong, X. & Harlow, D. (2015). Bulk Locality and Quantum Error Correction in AdS/CFT. *J. High Energ. Phys.* 1504:163.

Almheiri, A., Marolf, D., Polchinski, J. & Sully, J. (2013). Black holes: Complementarity or firewalls? *J. High Energ. Phys.* 2013:62.

Ammon, M., & Erdmenger, J. (2015). Strongly Coupled Condensed Matter Systems. In *Gauge/Gravity Duality: Foundations and Applications*. 460–504. Cambridge UK: Cambridge University Press.

Amoretti, A., Braggio, A., Maggiore, N. & Magnoli, N. (2017). Thermo-electric transport in gauge/gravity models. *Adv. in Physics: X* 2(2):409–27.

Bekenstein, J.D. (1973). Black holes and entropy. *Phys. Rev. D* 7:2333.

Beny, C., Kempf, A. & Kribs, D.W. (2007). Quantum error correction of observables. *Phys. Rev. A* 76:042303.

Bernstein, E. & Vazirani, U. (1997). Quantum complexity theory. *SIAM J. Comput.* 26(5):1411–73.

Bigazzi, B., Cotrone, A.L. & Evans, N. (2016). Holographic methods for strongly coupled systems. *Il Colle di Galileo*, S.l. 5(1):71–6.

Cleve, R., Gottesman, D. & Lo, H.-K. (1999). How to share a quantum secret. *Phys. Rev. Lett.* 83:648–51.

de Sitter, W. (1917). On Einstein's theory of gravitation, and its astronomical consequences. Third paper. *Mon. Not. R. Astr. Soc.* 78:3–28.

Dong, X., Silverstein, E. & Torroba, G. (2018). De Sitter holography and entanglement entropy. *J. High Energ. Phys.* 2018:50.

Eisert, J., Cramer, M. & Plenio, M.B. (2010). Area laws for the entanglement entropy: Review. *Rev. Mod. Phys.* 82:277.

Fruchtman, A., Lambert, N. & Gauger, E.M. (2016). When do perturbative approaches accurately capture the dynamics of complex quantum systems? Nature: *Sci. Rep.* 6(28204).

Hamilton, A., Kabat, D., Lifschytz, G. & Lowe, D.A. (2006). Local bulk operators in AdS/CFT: A boundary view of horizons and locality. *Phys. Rev. D* 73:086003.

Harlow, D. & Hayden, P. (2013). Quantum computation vs. firewalls. *J. High Energ. Phys.* 2013:85.

Harlow, D. (2017). The Ryu–Takayanagi formula from quantum error correction. *Commun. Math. Phys.* 354(865).

Hartnoll, S.A., Lucas, A. & Sachdev, S. (2018). *Holographic Quantum Matter*. Cambridge MA: MIT Press.

Hawking, S.W. (1975). Particle creation by black holes. *Commun. Math. Phys.* 43(3):199–220.

Hawking, S.W. (2005). Information loss in black holes. *Phys. Rev. D* 72(8): 084013.

Li, S. & Yee, H.-U. (2018). Relaxation times for chiral transport phenomena and spin polarization in strongly coupled plasma. *Phys. Rev. D* 98:056018.

Lopez-Arcos, C., Murugan, J. & Nastase, H. (2016). Nonrelativistic limit of the abelianized ABJM model and the ADS/CMT correspondence. *J. High Energ. Phys.* 2016:165.

Maldacena, J. (1998). The large N limit of superconformal field theories and supergravity. *Adv. Theor. Math. Phys.* 2:231–52.

Maldacena, J. (2012). The gauge/gravity duality. In: G.T. Horowitz (ed.), *Black Holes in Higher Dimensions*. Cambridge UK: Cambridge University Press, pp. 325–47.

Natsuume, M. (2015). AdS/CFT Duality User Guide. Springer. Lecture Notes in Physics. 903.

Ogushi, S. (2001). Conformal anomaly via AdS/CFT duality. *Soryushiron Kenkyu*. 103:6–84.

Osborne, T.J. & Stiegemann, D.E. (2017). Dynamics for holographic codes. arXiv:1706.08823 [quant-ph].

Pastawski, F., Yoshida, B., Harlow, D. & Preskill, J. (2015). Holographic quantum error-correcting codes: Toy models for the bulk/boundary correspondence. *J. High Energ. Phys*. 6(149):1–53.

Ryu, S. & Takayanagi, T. (2006). Holographic derivation of entanglement entropy from AdS/CFT. *Phys. Rev. Lett*. 96:181602.

Sachdev, S. (2011). Condensed Matter and AdS/CFT. In: E. Papantonopoulos (ed). *From Gravity to Thermal Gauge Theories: The AdS/CFT Correspondence. Lecture Notes in Physics*, Vol 828. Heidelberg, Germany: Springer.

Sergio, A. & Pires, T. (2014). *AdS/CFT Correspondence in Condensed Matter*. San Rafael, CA: IOP Concise Physics Morgan & Claypool Publishers.

Strydom, M. (2014). Magnetic Vortices in Gauge/Gravity Duality PhD Thesis. Munich, Germany. Ludwig Maximilians University.

Susskind, L. (1995). The world as a hologram. *J. Math. Phys*. 36(11):6377–96.

Swingle, B. (2012). Entanglement renormalization and holography. *Phys. Rev. D* 86:065007.

Vidal, G. (2008). A class of quantum many-body states that can be efficiently simulated. *Phys. Rev. Lett*. 101:110501.

Watrous, J. (2002). Quantum statistical zero-knowledge. arXiv:quant-ph/0202111.

Witten, E. (2016). An SYK-Like model without disorder. arXiv:1610.09758 [hep-th].

Zaanen, J., Liu, Y., Sun, Y.-W. & Schalm, K. (2016). *Holographic Duality in Condensed Matter Physics*. Cambridge, UK: Cambridge University Press.

Chapter 14

Holographic Quantum Error-Correcting Codes

Abstract

This chapter continues the research discussion that interprets the anti-de Sitter/conformal field theory (AdS/CFT) correspondence as an information-theoretic domain that (among other operations) can be error-corrected with quantum error-correction codes. Harlow & Hayden (2013) identify the AdS/CFT correspondence as an information theory problem, which implies computational complexity as an analysis tool, and the quantum information properties of entanglement and error correction. Almheiri *et al.* (2015) propose the interpretation of the AdS/CFT correspondence as a quantum error-correcting code. This model is realized by Pastawski *et al.* (2015), by enumerating a specific quantum error-correcting code. Such holographic codes may be used in a variety of applications, including in the development of technophysics-based quantum error-correcting codes.

14.1 Holographic Quantum Error-Correcting Codes

Holographic codes are proposed as specific quantum error-correcting codes derived from the principles of the anti-de Sitter/conformal field theory (AdS/CFT) correspondence.

14.1.1 *Quantum error correction*

Quantum error correction is a procedure used such that quantum information bits do not get lost or damaged. Quantum information bits are more sensitive to environmental damage and decay than their counterpart, classical information bits. The idea is to embed the qubits into a bigger state with more qubits so that even if some qubits are lost, the message can still be recovered (i.e. embed a smaller number of qubits into a large qubit state). A simple example of quantum error correction involves qutrits, as opposed to qubits. Qutrits are quantum information bits with three states, qubits are quantum information bits with two states (qu-trits: three, qu-bits: two). Ququats (four-state quantum information bits) have likewise been proposed (Traina, 2007). There is also the general notion of qudits (quantum digits or quantum bits with d-superpositions, units of quantum information described by a superposition of d states), for which Bell pair entanglement (nonlocality) has been tested in a 7-qudit system (Fonseca *et al.*, 2018). The idea is that quantum information bits are "artificial atoms", and as such, can be defined with arbitrarily many states; qubits (with two ending states, 0 and 1) are most aligned with the current information infrastructure based on classical bits (also with 0 and 1 values). The reason to use qutrits and other quantum information unit formats is that by having more states, they may be more efficient, for all operations, including error correction.

The 3-qutrit example is used to demonstrate quantum error correction. The model designates a larger Hilbert space and a smaller subspace. An overall system with 3-qutrits implies a 27-dimensional Hilbert space as its total scope of operation (3 qutrits $\times$ 3 states each = 27). The Hilbert space (or density matrix) of possible qutrit states is 27. A 3-qutrit code can be written to pick out a 3D subspace within the overall 27-dimensional Hilbert space. The implication for error-correcting is that because the qutrits are entangled, any 2 qutrits can be used to reconstitute a particular state. Since any 2 qutrits can be used for the reconstitution, this is called a unitary transformation, which implies symmetry (i.e. well-formedness) as a property of the system. The system protects against the erasure of any single qutrit because the state can be reconstituted from the other 2 qutrits.

This is a quantum error-correction scheme because any 2 qutrits can be used to recapitulate a particular system state. Likewise, in the AdS/CFT correspondence, it is possible to define bulk/boundary subregions to be related in an error-correction scheme. A smaller code subspace is defined in the bulk that is linked to a bigger Hilbert space in the boundary. Entangled physical qubits in the boundary can be used to recapitulate the logical qubit in the bulk. The general principle of quantum error correction at work in this situation is that if a subspace can be defined within a larger space, error correction may be possible. A smaller code subspace is defined within a larger space, from which systems states can be encoded and error corrected if necessary. The key point is that in a bulk/boundary correspondence, it is always possible to define a subspace of a bulk region and perform error correction through the entanglement relationship with the boundary region. The implication is the general claim is that any quantum mechanical subspace is a code space that is entangled and thus can be error-corrected. The error-correction protocol can be implemented in different ways for various purposes, for example to recover states, articulate the action of operators, reconstruct the structure of bulk locality, or attain other objectives.

14.1.2 *Tensor networks and MERA tensor networks*

Tensor networks are mathematical modeling tools for quantum many-body systems. Such tensor networks are collections of tensors with indices connected according to a computation graph or network pattern. The key advance of tensor networks is that they can be used to efficiently represent quantum many-body states with only polynomially many parameters (e.g. in a computable fashion). For example, specific physical quantities of interest in quantum many-body problems (such as correlation functions and local expectation values) can be reduced to the operation of contracting the indices of a tensor network.

Tensor networks have allowed tremendous progress in the domain of previously unsolvable quantum many-body problems. Quantum many-problems refer to the challenge that systems with more than even just two particles (i.e. quantum many-body systems) are nearly computationally impossible to calculate given the multi-dimensional particle movements of each quantum body and their interactions. Exact calculations of particle interactions are impractical for several reasons including the emergence of many-body effects as a result of electron correlations (each electron being influenced by the presence of other electrons). Tensor networks were introduced in the context of cerebellar neural networks, as a geometrization of brain function using tensors (Pellionisz & Llinas, 1980), and have been taken up in many fields, including in physics in relation to the holographic principle and the AdS/CFT correspondence (Maldecena, 1998; Witten, 1998). Tensor networks are used in the contemporary context of quantum computing to reduce dimensionality in high-dimensional tensor spaces, for tasks such as validating quantum-classical programming models with tensor network simulations (McCaskey *et al.*, 2018).

MERA is a tensor network method which specifically incorporates the entanglement property of quantum many-body systems. MERA (multiscale entanglement renormalization ansatz (guess)) combines tensor networks with entanglement renormalization (portability across scales) to create a family of tensor networks that efficiently approximates the wave functions of quantum bodies with long-range entanglement (Vidal, 2008). MERA provides an efficient representation of quantum many-body states on a d-dimensional lattice that is equivalent to a quantum circuit, and thus can be instantiated to compute exact measures such as local expectation values. The research advance is that the AdS/CFT correspondence can be modeled by a MERA-like tensor network in which quantum entanglement in the boundary theory is used as a building block for emergent bulk structure such as geometry (Swingle, 2012). The practical benefit is that MERA tensor network models can be used to prepare initial states of complicated quantum many-body systems for use in quantum computing and mathematical theory-testing. Hence, Pastawski *et al.* (2015) use a MERA tensor network instantiation to formalize the holographic quantum error-correction proposal made by Almheiri *et al.* (2015).

14.1.3 *AdS/CFT holographic quantum error-correcting codes*

Pastawski *et al.* (2015) implement the quantum error-correction interpretation of AdS/CFT proposed by Almheiri *et al.* (2015) with an explicitly solvable tensor network solution. A specific holographic code for quantum error correction is developed. The code is a tensor network instantiation of an isometric mapping between the bulk and the boundary that can be applied to prevent or correct errors. It is a quantum error-correction code between a logical qubit in the bulk that needs to be protected, and the ancilla of physical qubits in the boundary that protect it. The basic principle of quantum error correction is that information can be encoded into the long-range correlations of entangled quantum many-body states in such a way that it cannot be accessed locally. This is instantiated in the AdS/CFT correspondence model by protecting a logical qubit by entangling it with a bigger ancilla of physical qubits.

Pastawski *et al.* model the AdS/CFT correspondence with a MERA-type tensor network. A class of tensors called perfect tensors is used (perfect tensors are associated with pure quantum states of many spins that are maximally entangled). Using the perfect tensor network model, holographic states and codes are constructed. Holographic refers to the duality between the bulk and the boundary in the AdS/CFT correspondence. A *holographic state* is a mapping between a state of the system in the bulk, and a corresponding state of the system in the boundary. A *holographic code* is a quantum error-correction code that can be used to correct or prevent errors in a specified domain of the bulk/boundary correspondence.

The holographic code is enacted through the relationship between a bulk operator and a boundary operator. The idea is that the relationship between the bulk operator and the boundary operator functions as a quantum error-correction code. The relation between the two operators is defined by a mapping or dictionary. The tensor network is used to create a multi-dimensional linear model starting with one spin and expanding to the many spins in the system (this implements the AdS/CFT correspondence in a numerically analytical system that can be calculated). The perfect tensors provide an isometric encoding map of a specific quantum error-correcting code. The quantum error-correcting code encodes a single logical spin (in the bulk) in a larger block of physical spins (on the boundary), such that the logical spin is protected against erasure.

14.1.3.1 *Pastawski's holographic code*

Pastawski at al.'s quantum error-correcting code scheme consists of encoding a single logical spin in a block of ($2n$–1) physical spins, such that the logical spin is protected against the erasure of any (n–1) physical spins (Pastawski *et al.*, 2015, p. 5). Further, the quantum error-correcting code is the basis for a threshold-based quantum secret-sharing scheme. The code states have the property that a party holding any (n–1) qubits does not have information about the logical qubit, whereas a party holding any n-qubits has complete information about the logical spin (since erasure of the remaining (n–1) qubits is correctable). The structure of such secret-sharing schemes is a known cryptographic specification (Cleve *et al.*, 1999).

Pastawski *et al.* propose a 5-qubit code called the holographic pentagon code. The 5-qubit code is encoded as a six-leg tensor. The code is represented as a pentagon shape (five-sided) with six tensor legs. A network of the pentagon shapes with six-leg tensors is tiled out to cover a hyperbolic space (Pastawski *et al.*, 2015, p. 6). The holographic quantum code is represented as a network of such pentagons with six-leg tensors covering a hyperbolic space. Each pentagon has six tensor legs, one towards the bulk, two towards the neighbors on the left and right, and three towards the boundary. Each tensor has one open leg. This creates a structure for flowing computationally from the bulk to the boundary.

The open legs on the bulk are interpreted as the logical input legs of a quantum error-correcting code, and the open legs on the boundary are identified as outputs in which quantum information is encoded. The tensors have two types of indices, those that face the bulk and those that face the boundary. The indices on the legs are differentially contracted and uncontracted, with more contraction in the bulk logical space and less contraction in the boundary physical space. Overall, each tensor has at most two incoming indices from the previous layer and at most one bulk index. This provides an isometric mapping from incoming bulk indices to outgoing boundary indices and allows a computable dictionary to be specified between the bulk and the boundary. The entire tensor network can be viewed as an encoding isometry between the bulk and the boundary regions.

Through the holographic code, a system is created such that there are logical qubits in the bulk that need to be protected, and physical qubits on the boundary that comprise the ancilla to do the protecting. The whole structure effectively comprises a quantum error-correction code. Logical operators in the bulk can be mapped to physical operators in the boundary. The "pentagon-tiling tensor network is an isometric tensor from the bulk to the boundary", such that "we can view this isometry as the encoding transformation of a quantum error-correcting code" (Pastawski *et al.*, 2015, p. 6).

14.1.3.2 *Holographic codes: Qutrits and qubits*

The most basic quantum error-correction code is a stabilizer code which acts to flip the sign back in a quantum information bit that has been changed (i.e. stabilize the qubit). (Other quantum error-correction codes perform more sophisticated operations such as projection.) Almheiri *et al.* (2015) discuss a theoretical example of a quantum error-correction code with qutrits (three-state quantum information bits). The qutrit code is a [3,1,2]3 stabilizer code, in the format of [ancilla size, logical bit size, distance] bit format. This means having 3 physical qutrits, to protect 1 logical qutrit, over a distance metric of 2 (deletions up to 1 can be protected (2–1 = 1)), and the bit format is 3 (a qutrit, not a qubit). In the qutrit code, there is 1 logical quantum information bit distributed into 3 physical qutrits, each of which has 3 states, for a total of 9 elements (3 × 3). The 9-element qutrit code is conceptually similar to Shor's code which is qubit-based but also has 9 elements (Shor, 1995). This compares with the slightly shorter [7,1,3]2 Steane code, also qubit-based, which has a 7-qubit ancilla (Steane, 1996).

Whereas the qutrit example is more theoretical, Pastawski *et al.*'s holographic pentagon code is an explicit tensor network model that can be solved and implemented. The holographic pentagon code is a [5,1,3]2 code, meaning having 5 physical qubits, 1 logical qubit, with distance 3, and bit format 2 (qubit as opposed to qutrit). There is 1 logical degree-of-freedom in the bulk (at the center of the pentagon) related to 5 physical degrees-of-freedom on the boundary (the five sides of the pentagon). The distance is 3, meaning that information recovery is possible for up to two

Table 14.1. Examples of holographic quantum error-correcting codes.

Holographic code	Code structure	Quantum information bits		Distance and number of deletions protected	Quantum information bit format
		Physical	Logical		
Qutrit code	[3,1,2]3	3	1	2; 1	Qutrit
Pentagon code	[5,1,3]2	5	1	3; 2	Qubit

deletions (e.g. 3–1 = 2). The holographic code formats of the different methods are compared in Table 14.1.

14.1.3.3 *Practical error correction in quantum computing*

The benefit of holographic codes is that they represent a concrete possibility for quantum error correction in quantum computing that might be realized immediately on near-term NISQ devices. The earliest theoretical quantum error-correction codes were proposed by Shor (1995) and Steane (1996), and have been elaborated into practical implementations by other researchers such as Gottesman (2009). Pastawski's holographic code generalizes the concatenated quantum codes that Gottesman discusses for fault-tolerant quantum computing. Some of the next steps could include developing a theory of holographic codes that studies efficient schemes for correcting general errors beyond erasure errors, addressing problems such as the friction between correction rate and distance, and identifying ways to realize a universal set of logical operations acting on the codespace.

The benefit of the holographic model is that it provides a way to balance the trade-offs in quantum error-correction code design between redundancy and security, a familiar concern in network security design. On the one hand, quantum error-correction codes must be redundant enough to error-correct by having a certain number of ancilla qubits. On the other hand, quantum error-correction codes must be secure from external eavesdroppers. The threshold secret-sharing scheme property of the holographic code can be used to manage the trade-off between redundancy and security, by specifying the required threshold for secure codes and efficiently sized qubit ancillas.

14.1.3.4 *Causal and entanglement wedge reconstructions*

Pastawski *et al.* recapitulate the Rindler causal wedge reconstruction, and make additional progress toward the entanglement wedge reconstruction in the AdS/CFT correspondence discussed by Almheiri *et al.* (2015). The different wedge reconstructions are important because they indicate the range over which error correction can be conducted through the subregion–subregion correspondence in the overall the bulk/boundary relationship. The causal wedge reconstruction of bulk logical operators is implemented in the tensor network model. A Pauli operator (which manages quantum mechanical spins) is injected into an arbitrary open tensor leg in the bulk. The operator is defined such that it can be pushed into three additional legs of the tensor, which are in turn injected into neighboring tensors. By repeatedly pushing operators to the boundary of the network, eventually there can be some representation of the operator in the boundary region. The bulk operator is contained inside the causal wedge of this boundary region, thereby demonstrating the causal wedge reconstruction. Different operators can be pushed into the boundary by choosing different tensor legs, which leads to different representations of a logical operator.

The more expansive entanglement wedge reconstruction is difficult to calculate due to aspects of the Ryu–Takayanagi formula (2006), in which boundary entanglement entropy is related to the area of a corresponding minimal surface in the bulk (in particular, it is difficult to determine the entropy of the boundary state). However, different kinds of entanglement wedges might be tested such as the greedy entanglement wedge and the geometric entanglement wedge (Pastawski *et al.*, 2015, 33). Subsequent work has established a workable entanglement wedge reconstruction in the AdS/CFT correspondence (Dong *et al.*, 2016). The benefit of the entanglement wedge approach is that it offers a greater range over which the AdS/CFT correspondence can be formally defined, as compared with the smaller range of the causal wedge approach.

14.1.3.5 *Implications of the holographic code*

Overall, Pastawski *et al.* demonstrate that holographic codes are able to provide a concrete realization of the AdS/CFT correspondence. There is a quantitative confirmation of the quantum error-correction interpretation of the AdS/CFT correspondence. This is achieved by defining code subspaces, deriving a holographic error-correcting code scheme, and performing quantum error correction on information in the code subspaces. The quantum error correction is executed by using the bulk reconstruction to define a code subspace, entangling this code subspace with a reference system (a dictionary mapping between the bulk and the boundary), preparing a state with a MERA tensor network, and determining how the information operates between the two domains such that it can be error corrected. In the MERA constructions, tensors act on both on the boundary and bulk through the reference dictionary. The biggest potential use of these holographic codes could be in quantum computing, for quantum error correction in near-term NISQ devices.

14.2 Other Holographic Quantum Error-Correcting Codes

A range of other research projects take up the holographic correspondence and related tensor network constructions to build different kinds of quantum error-correcting codes and models of the bulk/boundary correspondence. Pastawski *et al.*'s holographic quantum error-correcting code is a code that is an isometric mapping of qubits (two-state quantum information units) between the bulk and the boundary (2015). The method is symmetric (six-legged tensors whose indices allow for mathematical manipulation), and relies on perfect tensors to generate maximal entanglement (the entanglement of qubits that is necessary to perform the quantum error correction).

Another approach specifies a holographic code using a 4-qutrit state to produce maximal entanglement (Latorre & Sierra, 2015). The slightly larger holographic code (4 qutrits × 3 states each) is made of a superposition of states whose relative distances are exponentially large with the size of the boundary. The benefit of a larger code operating on a longer distance suggests that the holographic code could be a candidate for quantum memory. Holographic codes might be used for different purposes, shorter codes for efficient error correction and longer codes for quantum memory. Another more complicated approach that might accommodate quantum memory is one that manages spins with Kraus operators (involving operator sums) (Verlinde & Verlinde, 2013), as opposed to the single Pauli operators employed by Pastawski *et al.* Also regarding quantum memory, other research uses the holographic correspondence to probe the theoretical limits of information storage in quantum systems by examining quantum error-correcting codes based on correlated spin phases and other system features (Yoshida, 2013).

Whereas quantum memory implies longer codes, the other research trajectory continues to discover shorter codes for more efficient error correction. One project proposes shorter codes with shorter distance, using tensor formulations to find quantum stabilizer codes that are more efficient because they require less distance (Bacon *et al.* 2017). Another approach elaborates more efficient codes by shrinking the operators in the holographic correspondence (Hirai, 2019).

14.2.1 *Emergent bulk geometry from boundary entanglement*

Pastawski *et al.*'s approach engages maximal entropy through perfect tensors in order to generate a quantum error-correcting code, and develops an isometric mapping between the bulk and the boundary. Another approach focused on cosmology research does not need entanglement for error correction and employs a more straightforward tensor tree model (hierarchical tensor trees) with an exact unitary mapping between the bulk and the boundary (Qi, 2013). The work examines an emblematic topic of interest in AdS/CFT correspondence research, how structure in the bulk (geometry) arises from the relation with boundary entanglement. The project investigates the emergent bulk geometry corresponding to different boundary states, by testing different scenarios involving mass and temperature.

Whereas Qi uses a tensor-tree model with an exact unitary bulk/boundary mapping, Pastawski *et al.* use a perfect tensor model with an isometric bulk–boundary mapping. The benefit of the isometric mapping over the exact unitary mapping is that it is a more flexible model for the mapping, instead of an exact relationship. This gives the isometric mapping model the extensibility to accommodate additional degrees of freedom by expanding and contracting the tensor leg indices (including by adding additional parameters to the indices, for example, a time parameter). Pastawski's isometric model is an expandable structure for general purpose quantum error correction, whereas Qi's exact unitary tensor-tree model is designed to solve targeted problems.

14.2.2 *Ryu–Takayanagi quantum error correction codes*

Performing quantum error correction between the bulk and the boundary regions requires articulating how entropy in the two regions operates differently. Whereas bulk entropy scales by area, boundary entropy scales by volume, and their interrelation is non-trivial. The Ryu–Takayanagi formula for entanglement entropy links the two, by relating the entanglement entropy of a boundary region to the area of a corresponding minimal surface in the bulk (Ryu & Takayanagi, 2006). Research continues on the issues raised in this body of work. A quantum error-correction interpretation for the Ryu–Takayanagi formula for entanglement entropy is proposed (Harlow, 2017). The quantum error-correction interpretation is based on the specification of a new subalgebra code. The subalgebra code (subalgebra: closed under all of its operations) produces a complementary recovery mechanism between the boundary entanglement and the bulk geometry, using the entanglement wedge reconstruction method. The subalgebra code is intended as a general purpose technology that might be used as a standard feature in any well-formed quantum error-correction code that is based on the holographic correspondence.

14.2.3 *Extending MERA tensor network models*

Other work extends tensor networks and MERA-type models, specifically as used in the context of quantum error correction (Ferris & Poulin, 2014). An important advance is implementing the renormalization group in a MERA tensor network environment (Evenbly & Vidal, 2015). Renormalization is a mathematical technique that allows a system to be investigated at different scales, especially as it changes dynamically and undergoes phase transition. The work is based on the renormalization group (Wilson, 1971). The initial renormalization group articulates scaling near the critical point for an Ising ferromagnet in a differential form.

The renormalization group helps to address the so-called Kondo problem of understanding magnetic impurities in non-magnetic metals (the impurity problem is related to the interaction between localized magnetic impurities and wandering electrons) (Kondo, 1964), and also the problem of infinities arising in the application of quantum field theories. The renormalization group has been established as a canonical analysis tool for phase transition (Goldenfeld, 1992).

Impurities in materials (natural or doped) are a form of anomaly that can be exploited in analytical models to test the robustness and invariance of the system. Evenbly & Vidal employ two quantum critical Hamiltonians to test between two phase transition situations. The theoretical result is the notion of using directed influence in the renormalization group (closely following Wilson). The practical result is that directed influence is a method that can be implemented in tensor networks to exploit impurities (anomalies) in otherwise homogeneous systems. Other research uses the renormalization group to investigate the strong-coupling limit (the moment of exchange between light and matter), which is an important feature of superconductive materials. The renormalization group calculations reveal anomalies in the strong-coupling limit, such as impurity entropy and the effective impurity moment (Fritz, 2006). Some of these kinds of insights gained from the application of the renormalization group in MERA tensor networks might lead to improved superconducting materials and error-correction methods for quantum computing.

14.2.4 *Bosonic error-correction codes*

Beyond basic quantum error-correction codes, more efficient quantum error-correction codes have been proposed which take greater advantage of quantum properties, notably bosonic error-correcting codes. One way of distinguishing subatomic particles is between the two classes of bosons and fermions. Whereas bosons like to clump together, fermions do not. Particles that obey the Pauli Exclusion Principle are called fermions, and those that do not are called bosons. Basically, the Pauli Exclusion Principle means that some particles are exclusive in that they do not like to clump together (fermions), whereas others do (bosons). Fermions do not like to be together in the same space (and do obey the Pauli Exclusion Principle), bosons like to share space (and do not obey the Pauli Exclusion Principle). The important effect of the two kinds of clumping (bosons) and non-clumping (fermions) subatomic particles is that this gives rise to chemistry.

Boson (particle-clumping) quantum error-correction codes make use of the feature that bosons like to clump together which suggests that the error-correction mechanism can be spread out efficiently on the clumped particles. Different quantum error-correction schemes have been proposed based on different bosonic modes. Bosonic modes refer to any of several specific states of the boson related to its spin, charge, polarization, or wave form. The basic classes of bosonic error-correcting codes are cat codes and binomial codes which are based on the superposition of states written in different ways, such as by phase-space rotation symmetry or with binomial coefficients. In particular, binomial codes may be able to exactly correct errors that are polynomial up to a specific degree in the creation of bosons and annihilation operators, including correcting for situations of both amplitude damping and displacement noise (Michael *et al.*, 2016).

Other approaches to bosonic error-correction codes are also proposed. One idea is storing error-correctable quantum information in bosonic modes, engaging the property that bosons are natural oscillators. The benefit is that an oscillator has infinitely many energy levels in a single physical system. Hence, an error-correction scheme might be more efficient with codes that take advantage of the redundancy within a single system, as opposed to duplicating many two-level qubits. For instance, an oscillator could be encoded into many oscillators (Noh *et al.*, 2019). Bosonic codes are often instantiated in a multi-dimensional space called the Fock basis. The Fock basis (starting configuration), or Fock space, is an algebraic construction of space that is used in quantum mechanics to construct the quantum state space of a variable or unknown number of identical particles from a single particle Hilbert space.

14.3 Quantum Coding Theory

Quantum coding theory is an emerging area in quantum information science that studies the development and use of various forms of quantum classical error-correcting codes (low-density parity-check, LDPC, codes). Then the performance of the codes is analyzed using a belief propagation decoding algorithm.

The advance of the proposed error-correction solution is using an error-correction-type architecture for novel purposes beyond the initial use in error correction. The method takes advantage of having a smeared out ancilla of extra entangled qubits as a quantum system feature, and deploys this in new ways, such as for local controllability in a quantum annealing solver. Error correction could become a standard feature of quantum smart networks, for the immediate purpose of quantum error correction, and including a wider set of potential uses. The overall concept is improved system manipulation and control through quantum error-correction-type models.

References

Almheiri, A., Dong, X. & Harlow, D. (2015). Bulk locality and quantum error correction in AdS/CFT. *J. High Energ. Phys.* 163:1–33.

Bacon, D., Flammia, S.T., Harrow, A.W. & Shi, J. (2017). Sparse quantum codes from quantum circuits. *IEEE Trans. Inf. Theory* 63(4):2464–79.

Cleve, R., Gottesman, D. & Lo, H.-K. (1999). How to share a quantum secret. *Phys. Rev. Lett.* 83:648–51.

Dong, X., Harlow, D. & Wall, A.C. (2016). Reconstruction of bulk operators within the entanglement wedge in gauge-gravity duality. *Phys. Rev. Lett.* 117:021601.

Evenbly, G. & Vidal, G. (2015). A theory of minimal updates in holography. *Phys. Rev. B* 91:205119.

Ferris, A.J. & Poulin, D. (2014). Tensor networks and quantum error correction. *Phys. Rev. Lett.* 113:030501.

Fonseca, A., Rosier, A., Vertesi, T. *et al.* (2018). Survey on the Bell nonlocality of a pair of entangled qudits. *Phys. Rev. A* 98:042105.

Freedman, M. & Headrick, M. (2017). Bit threads and holographic entanglement. *Comm. Math. Phys.* 352(407).

Fritz, L. (2006). Quantum Phase Transitions in Models of Magnetic Impurities. PhD Thesis: Physics, Karlsruhe University.

Goldenfeld, N. (1992). *Lectures on Phase Transitions and the Renormalization Group*. Boulder, CO: Westview Press.

Table 14.2. Quantum information science topics.

1.	Transmitting classical information over quantum channels
2.	Transmitting quantum information over quantum channels
3.	Considering the trade-offs between acquiring information about a quantum state and disturbing the state
4.	Quantifying quantum entanglement

codes. Quantum information science more generally is a new frontier of physical science in which concepts related to quantum information processing such as entanglement and quantum error correction are proving relevant to a wide range of contemporary problems. The field of quantum information science addresses four main kinds of topics as enumerated in Table 14.2. These include the transmission of classical information over quantum channels, the transmission of quantum information over quantum channels, the trade-offs between acquiring information about a quantum state and disturbing the state, and quantifying quantum entanglement (Preskill, 2015).

There are several immediate practical applications for quantum coding theory. One is quantum error-correcting codes in quantum computing. Another application is quantum stabilizer codes in secure end-to-end communication in the quantum internet (Wilde, 2008). A third application is the proposal of using quantum error-correcting codes in condensed matter physics as quantum memory. There could be active quantum error correction on topological codes for long-term qubit storage as a potential model for quantum memory (Lang & Buchler, 2018). A related method suggests finding ways to dynamically error-correct topological phases of matter such that topological quantum memories might be established (Zeng *et al.*, 2018).

Theoretical applications of quantum coding theory are indicated in the same sense that quantum information science is emerging more generally as a tool for sharpening the understanding of various problems in quantum mechanics and beyond. An important result of the AdS/CFT correspondence is demonstrating the value of information-theoretic approaches to physics. Likewise, quantum coding theory and quantum error correction might be general methods with wide applicability.

14.4 Technophysics: AdS/Deep Learning Correspondence

There is a growing body of technophysics work in AdS/deep learning (AdS/DL) correspondence models. One project presents a deep learning interpretation of the AdS/CFT correspondence which describes the emergence of bulk structure (Hashimoto *et al.*, 2018). The advent of a bulk metric function is demonstrated through a learning process applied to data sets given in boundary quantum field theories. The emergent radial direction (outward-moving) of the bulk is identified with the depth of the layers, and the network itself is interpreted as a bulk geometry. The deep learning network provides a data-driven holographic model of strongly coupled systems. In terms of a scalar theory with unknown mass and coupling, in unknown curved spacetime with a black hole horizon, the deep learning network is used to specify a framework that fits the given data. With the experimental quantum field theory data as input, the neural network is able to determine the bulk metric, the mass, and the quadratic coupling of the holographic model. Hence, the AdS/DL correspondence might be a useful tool for modeling gravitational systems and other strongly correlated systems.

Other technophysics research in the area of the AdS/deep learning correspondence also connects the AdS/CFT correspondence to network science and deep learning methods (Freedman & Headrick, 2017). A theorem from network theory, the max-flow–min-cut principle (related to partition functions) is used. The max-flow–min-cut theorem has been widely applied to networks and Riemannian manifolds, and thus might be applied similarly to the holographic correspondence. The Ryu–Takayanagi entropy entanglement formula is addressed, which relates the entanglement entropy of a boundary region to structure in the bulk, namely the area of a corresponding bulk minimal surface. The max-flow–min-cut theorem is used to rewrite the Ryu–Takayanagi formula. The Ryu–Takayanagi formula is rewritten not referencing the minimal surface, but instead invoking the notion of flux flowing across vector fields. The entanglement entropy of a boundary region is given by the maximum outgoing flux. The flux threads represent the entanglement between points on the boundary, which implements the holographic principle. The benefit of the max-flow–min-cut model is that it is a method for articulating entanglement entropy in the boundary, a measure which can be difficult to derive. The method may be used practically as well as theoretically to understand more about the properties of Ryu–Takayanagi entropy entanglement.

14.4.1 *Novel uses of quantum error-correction architecture*

In machine learning, quantum error correction has been proposed as a method for allowing greater control during quantum annealing processes (Lechner *et al.*, 2015). Quantum annealers are physical devices that attempt to solve optimization problems by exploiting quantum mechanics. A standard method of quantum annealing is encoding the optimization problem in the Ising interactions between qubits, in a spin glass model. Controlling the qubit interactions in the spin glass model is not possible, though, and the quantum computation must run until it is finished. However, instead of using a spin glass model of the Ising interactions, an error-correction model could be used to engage certain qubits selectively as they perform the computation, thus offering a control mechanism for the computation.

The proposed solution is in the form of a scalable architecture with fully controllable all-to-all connectivity that can be implemented selectively on local interactions. The method is in the form of a quantum error-correction scheme. The input of the optimization problem is encoded in local fields that act on an extended set of physical qubits. The output is redundantly encoded in the physical qubits, which instantiates an error-correction architecture for the system. The control mechanism is applied by allowing only certain portions of the output into the computation. The error-correction architecture creates a tableau of extra qubits that can be engaged selectively to perform the computation, from the spread-out entanglement of qubits in the error-correction scheme. More formally, the model can be understood as a lattice gauge theory, in which long-range interactions are mediated by gauge constraints.

Machine learning methods have been deployed to favorably analyze the effectiveness of these quantum error-correcting codes (Pastawski & Preskill, 2016). The quantum error-correcting codes are first interpreted as

Gottesman, D. (2009). An introduction to quantum error correction and fault-tolerant quantum computation. arXiv:0904.2557 [quant-ph].

Harlow, D. & Hayden, P. (2013). Quantum computation vs. firewalls. *J. High Energ. Phys.* 2013:85.

Harlow, D. (2017). The Ryu–Takayanagi formula from quantum error correction. *Commun. Math. Phys.* 354(865).

Hashimoto, K., Sugishita, S., Tanaka, A. & Tomiya, A. (2018). Deep learning and the AdS/CFT correspondence. *Phys. Rev. D* 98:046019.

Hirai, H. (2019). Shrinking of operators in quantum error correction and AdS/CFT. arXiv:1906.05501 [hep-th].

Kondo, J. (1964). Resistance minimum in dilute magnetic alloys. *Prog. Theor. Phys.* 32(1):37–49.

Lang, N. & Buchler, H.P. (2018). Strictly local one-dimensional topological quantum error correction with symmetry-constrained cellular automata. *SciPost Phys.* 4(007).

Latorre, J.I. & Sierra, G. (2015). Holographic codes. arXiv:1502.06618 [quant-ph].

Lechner, W., Hauke1, P. & Zoller, P. (2015). A quantum annealing architecture with all-to-all connectivity from local interactions. *Sci. Adv.* 1(9):e1500838.

Maldecena, J.M. (1998). The large N limit of superconformal field theories and supergravity. *Adv. Theor. Math. Phys.* 2(231).

McCaskey, A., Dumitrescu, E., Chen, M. *et al.* (2018). Validating quantum-classical programming models with tensor network simulations. *PLoS ONE* 13(12):e0206704.

Michael, M.H., Silveri, M., Brierley, R.T. *et al.*, (2016). New class of quantum error-correcting codes for a bosonic mode. *Phys. Rev. X* 6:031006.

Noh, K., Girvin, S.M. & Jiang, L. (2019). Encoding an oscillator into many oscillators. arXiv:1903.12615 [quant-ph].

Pastawski, F. & Preskill, J. (2016). Error correction for encoded quantum annealing. *Phys. Rev. A* 93(5):052325.

Pastawski, F., Yoshida, B., Harlow, D. & Preskill, J. (2015). Holographic quantum error-correcting codes: Toy models for the bulk/boundary correspondence. *J. High Energ. Phys.* 6(149):1–53.

Pellionisz, A. & Llinas, R. (1980). Tensorial Approach to the geometry of brain function: Cerebellar coordination via a metric tensor. *Neuroscience* 5(7):1125–36.

Preskill, J. (2015). *Lecture Notes for Ph219/CS219: Quantum Information and Computation*. California Institute of Technology.

Qi, X.-L. (2013). Exact holographic mapping and emergent space-time geometry. arXiv:1309.6282.

Ryu, S. & Takayanagi, T. (2006). Holographic derivation of entanglement entropy from AdS/CFT. *Phys. Rev. Lett.* 96:181602.

Shor, P. (1995). Scheme for reducing decoherence in quantum computer memory. *Phys. Rev. A* 52:R2493(R).

Steane, A. (1996). Simple quantum error correcting codes. *Phys. Rev. A* 54:4741.

Swingle, B. (2012). Entanglement renormalization and holography. *Phys. Rev. D* 86:065007.

Traina, P. (2007). Violation of local realism for ququats. *Open Syst. Inf. Dyn.* 14(2):217–22.

Verlinde, E. & Verlinde, H. (2013). Black hole entanglement and quantum error correction. *J. High Energ. Phys.* 2013:107.

Vidal, G. (2008). A class of quantum many-body states that can be efficiently simulated. *Phys. Rev. Lett.* 101:110501.

Wilde, M.M. (2008). Quantum Coding with Entanglement. PhD. Thesis. University of Southern California.

Wilson, K.G. (1971). Renormalization group and critical phenomena. I. renormalization group and the Kadanoff scaling picture. *Phys. Rev. B* 4:3174–83.

Witten, E. (1998). Anti de sitter space and holography. *Adv. Theor. Math. Phys.* 2(253).

Yoshida, B. (2013). Information storage capacity of discrete spin systems. *Annals Phys.* 338(134).

Zeng, B., Chen, X., Zhou, D.L. & Wen, X.G. (2018). Quantum information meets quantum matter: From quantum entanglement to topological phase in many-body systems. arXiv:1508.02595 [cond-mat.str-el].

Part 6

Quantum Smart Networks

Chapter 15

AdS/Smart Network Correspondence and Conclusion

Abstract

The smart network theories are derived from statistical physics (statistical neural field theory and spin glass models) and information theory (the the anti-de Sitter/conformal field theory, AdS/CFT, correspondence). Whereas the smart network field theory (SNFT) is aimed at the next phases of the classical development of smart network systems, the smart network quantum field theory (SNQFT) is intended to facilitate the potential expansion into the quantum domain, in the implementation of quantum smart networks. This chapter develops the SNQFT from the AdS/CFT correspondence, and suggests the possibility of the correspondence supplanting probability as a central organizing principle for certain ways of understanding physical reality. The SNQFT is developed and motivated towards a variety of potential applications. It is suggested that everyday macroscale reality is the boundary to the quantum mechanical bulk. An emphasis is placed on models that integrate quantum mechanical field views and geometric views, including for the exploitation of quantum mechanical systems in macroscale reality. Finally, the work considers risks and limitations and concludes. The content in this chapter is more speculative and conjectural than that in previous chapters.

15.1 Smart Network Quantum Field Theory

Smart network technologies are already quantum-ready, in the sense that they are instantiated in 3D formats. The 3D formats are computation graphs which imply programmability, analytic solvability, and some translation to quantum entanglement, interference, and superposition states in a Hilbert space. This is a good first step, however, there could be many other requirements for the realization of quantum smart networks (smart network technologies implemented in quantum computing environments). The smart network quantum field theory (SNQFT) aims to support these steps.

15.1.1 *AdS/CFT correspondence-motivated SNQFT*

The anti-de Sitter space/conformal field theory (AdS/CFT) correspondence (also called gauge/gravity duality) is selected for the derivation of the smart network quantum field theory (SNQFT) for several reasons. First, the main issue that the AdS/CFT correspondence addresses is likewise a key concern as smart networks potentially migrate to the quantum domain. This is the linking of two regions, one of a macro-level surface, and one of a more complicated higher-dimensional bulk. The AdS/CFT correspondence addresses the relation between the boundary and the more complex bulk region determined by gravity or geometrical aspects, including space and time. Smart networks too exist on a macro-level surface and may be extending into a quantum mechanical bulk. The key motivation is to create a model for linking the quantum and non-quantum domains such that the quantum realm can be activated in a useful way at the macroscale.

The second reason that the AdS/CFT correspondence is used to inspire the smart network quantum field theory is that it has a rich application lineage in other domains (including superconducting materials, condensed matter physics, topological matter, and plasma physics), which establishes a precedent for its wider application. There are already some technophysics applications of the AdS/CFT correspondence underway in deep learning and network theory, suggesting the validity and continued trajectory of the approach. From a practical perspective, there are tools such as MERA tensor networks and quantum error-correction codes for the analytic solvability of problems structured in the form of the correspondence.

Third, smart networks are complex systems and the AdS/CFT correspondence is a ComplexityTech (a technology for managing complexity). Information theory and computational complexity theory have been demonstrated to apply to the AdS/CFT correspondence, which implies that these techniques can be used to manage complex smart network systems. Further, the AdS/CFT correspondence exposes some of the inherent information-theoretic security properties of the quantum domain which could be important in quantum smart nctworks, and are already visible in contemporary smart network design (discussed in Section 15.1.3). Finally, from a complexity perspective, the Ads/CFT correspondence is a good candidate to inspire the SNQFT because it explains not only system characterization, dynamics, and criticality, but also serves as a model for identifying novel emergence within systems (the emergence of bulk structure).

15.1.2 *Minimal elements of smart network quantum field theory*

The minimal elements of the SNQFT are described in Table 15.1. These include the system structure, dynamics and operation, criticality, and a

Table 15.1. SNQFT: Minimal elements.

Element	Function
1. System structure	Particles (physical): atoms, ions, photons Particles (logical): qudits (qubits, etc.) Interactions: coupling, entanglement
2. System dynamics and operation	Quantum states, histories Quantum gate logic operations Error correction and computational result
3. System criticality	Quantum threshold trigger Quantum phase transition Quantum optimal control mechanism
4. System novelty	Emergence of bulk structure (geometric properties, space and time properties)

new category, novelty, connoting that the smart network quantum field theory allows for the emergence of novelty in smart network systems. The smart network quantum field theory may have a literal interpretation in quantum systems, and also analogical interpretations. Even without considering the smart network domain, qubits (any kind of quantum information units) are analogical, in the sense that they are "artificial atoms" used for logical computation.

The particles in quantum networks are atoms, ions, and photons. Particles have the special properties of quantum objects, which are superposition, entanglement, and interference (SEI) properties. The particle interactions are couplings, gate operations, and entanglement relationships. The system operation and dynamics comprise quantum states and histories of the system, and problem setup, computation, error correction, and result. System criticality is likewise important in quantum systems, with corresponding analogs for threshold triggers, phase transition, and optimal control mechanisms.

The particles are defined as any variety of qudits (quantum digits or quantum information bits with d-superpositions, units of quantum information described by a superposition of d states). The most frequently-used qudit is the qubit, a two-state quantum information unit (0/1) that is most closely related to the classical information bit (0/1). Qutrits (three-state quantum information bits) and ququats (four-state quantum information bits) are also proposed. As much as a 7-qudit system has been tested (Fonseca *et al.*, 2018).

One implication is that more efficient computing and error-correcting may be available by instantiating operations within a single system rather than by spreading them out over multiple computation units. It might be possible for a self-contained qudit system to do its own computation, error correction, and proof, as a complete unit of computational complexity. The possibility of greater latitude in defining the computation unit raises important theoretical questions about computational complexity and efficiency. There are questions about which kinds of core computational units and structures are best for certain tasks. All aspects of computation theory are open to questioning, including number bases; whether decimal (base 10), binary (base 2), or other expansions are most relevant.

15.1.2.1 *Novelty emergence*

Well-formed smart network theories must explain not only system characterization, dynamics, and criticality, but also novel emergence. The design objective of using the AdS/CFT correspondence is that it is a tool for describing emergence. The correspondence articulates how bulk locality emerges (local structure in the bulk), namely how geometry arises, and space and time properties. The way that geometry is related to space and time is through curvature (in general relativity, Riemannian curvature gives rise to space and time). The immediate practical benefit is that the bulk emergence of geometry, space, and time properties could be relevant for creating and understanding the next phases of smart network technology development. The ongoing theoretical benefit is that a mechanism for identifying and harnessing novel emergence is a valuable function that could further contribute to the understanding of any system, including smart network technologies.

15.1.3 *Nature's quantum security features*

15.1.3.1 *Black holes are zero-knowledge proofs*

Security features are among the first and most important class of applications for quantum smart networks since many classical systems are not quantum-secure. The AdS/CFT correspondence reveals some of the inherent security features of the quantum domain that could be important in quantum computing, and that are already emerging in smart network design. These include quantum error correction, quantum zero-knowledge proofs, and quantum statistics (Table 15.2). Black holes are zero-knowledge proofs in the sense that they perform their own information validation proofs as any quantum computational complexity domain.

15.1.3.2 *Enumerating nature's quantum security features*

The quantum world appears to have security features built into it naturally. One set of security features relates to the no-cloning theorem (quantum information cannot be copied) and the no-measurement principle (quantum information cannot be looked at or measured without changing or

Table 15.2. Natural security features built into quantum mechanical domains.

Principle	Security feature
1. No-cloning theorem	Cannot copy quantum information
2. No-measurement principle	Cannot measure quantum information without damaging it (eavesdropping is immediately detectable)
3. BQP/QSZK computational complexity	Quantum information performs its own computational verification (zero-knowledge proofs)
4. Quantum statistics	Provable randomness: distributions could only be quantum-generated
5. Quantum error correction	Error correction of bulk regions through boundary regions

damaging it, hence, eavesdropping is immediately detectable). Another class of security features concerns the quantum statistical zero-knowledge (QSZK) computational complexity class (quantum computing performs its own computational verification (zero-knowledge proof technology)). For this class of computing problems, the verification can be performed by the computer itself without the need for interactions with a prover.

Other security features are available through quantum statistics (certain quantum statistical distributions based on amplitude (interference), entanglement, or superposition that could only have been generated by quantum computers and thus convey provable randomness). Particles are always in the form of a distribution due to wave motion. Further quantum security features are provided through error correction, in the structure of the holographic correspondence. In the AdS/CFT correspondence, because a boundary theory in one fewer dimensions can describe a bulk theory, nature effectively performs its own error correction. The 2D surface recapitulates (error corrects or protects) the information in the 3D bulk. The most basic hologram (or black hole) can be seen as a quantum error correction device in this way.

15.1.3.3 *Deploying nature's quantum security features*

An overarching principle in technology design (including per Feynman on quantum computing) is that technology that is most closely aligned with the underlying phenomenon with which it is related may be the most expedient. Apparently, unintentionally, smart network technology design has already started to incorporate some of the natural security features of the quantum mechanical domain. Occam's razor may be steering these results (the most efficient solution is the best), unwittingly finding techniques that are congruent with nature. However, a more explicit technophysics approach could be invoked to deliberately integrate nature's suite of quantum mechanical security features into smart network technology.

The AdS/CFT correspondence indicates how nature's quantum security features might be engaged to make various claims about quantum smart network systems. First, the computational complexity class QSZK has been used to argue that calculating information about an outwardly-radiating qubit from a black hole would not be computable in polynomial time on a quantum computer. The zero-knowledge property is implicated, in that there is no traditional prover–verifier relationship because the quantum computer does the verification so fast that the prover is not necessary. The computational complexity class QSZK contains the class bounded-probability quantum polynomial (BQP), which is more generally the class of problems that can be solved with a quantum computer. For such problems, the verification can be conducted directly using the computer itself without the need to interact with a prover (Watrous, 2002). Smart networks are incorporating this concept too, in the sense that the computational verification capability of zero-knowledge proof technology likewise does not always require the involvement of the prover. The implication is that quantum computing has zero-knowledge proof technology built in as a feature: BQP, the class of problems solvable with quantum computers, performs its own verification.

In another way, zero-knowledge proofs are in the shape of a hologram. There are two different perspectives of the same information that both evaluate as true. Proofs are in the form of a two-level information system, the underlying information and the proof as an assessment of the underlying information. Both evaluate as true from different views of the information. The local observer (prover) sees the dimensional detail of the private information and knows the information is true, whereas the remote observer (verifier) sees the one-fewer dimensional evaluation of the information as a one-bit value that is true.

Nature's quantum security features are also seen in error correction. It is argued that quantum error-correction codes based on a threshold secret-sharing scheme might extend the potential range of error correction available in quantum systems (Almheiri *et al.*, 2015). The claim also improves the precision of the AdS/CFT correspondence between the bulk and the boundary regions. A specific holographic quantum error-correction code is proposed in the form of a threshold secret-sharing scheme. Threshold secret-sharing schemes are becoming a standard feature of smart networks too, in digital signatures, multi-party computing, and Layer 2 payment channels. The immediate application is that the formalized connection between quantum error correction and holography offers a model for implementing quantum error correction in quantum computers. This further supports the idea that a substantial degree of reliable quantum information processing might be achieved with imperfect physical components (NISQ devices).

15.1.4 *Random tensors: A graph is a field*

Beyond security features, the AdS/CFT correspondence implicates the potential for novel discovery, including through random tensor networks. An extension of tensor network models is random tensors (Gurau, 2016). Random refers to the process of calculating wave functions, which involves integrating over a random configuration of fields. Random tensors are used to generalize random matrices (fields) to higher dimensions. A field is a matrix, and a field theory is computable as a matrix model. The idea is to compute quantum mechanical systems by formulating them as a graph or matrix problem. In particular, random tensor network models are used to study the next scale level below quantum chromodynamics, namely gauge theories of subatomic particles involving gluons and quarks. The new element conveyed by random tensor networks is two things, mainly in their context of use, a $1/N$ expansion, but also an idea related to the AdS/CFT correspondence.

The so-called one over large-N expansion refers to a situation that arises in gauge theories. In gauge theories, the random tensor model is used to formulate random $N \times N$ matrices to support a perturbative expansion of the system in terms of graphs. Unlike in general quantum field theories, at the subatomic particle scale, a new parameter, large-N, appears in random tensor models. (This is somewhat similar to the way that infinities appear in quantum field theories, and are resolved with the renormalization group.) A key advance in random tensors is establishing a $1/N$ expansion for tensors to manage the appearance of large-N (Baratin & Oriti, 2010).

The key conceptual point of random tensors for the AdS/CFT correspondence (gauge/gravity duality) is that geometries are fundamental fields and that geometry is dynamical (Hayden *et al.*, 2016). By focusing at the scale of gauge theories, random tensors reveal the idea that geometry is a dynamical field. One implication is that if geometry is a field, it can be computed. Another point is that geometry is dynamical, it changes, and random tensors can be used to calculate this too. A third implication is that random tensor networks provide a way to further combine quantum mechanical theories (fields) and geometric theories, by summing over random configurations of dynamic geometrical fields.

Random tensors provide another analytical tool for the AdS/CFT correspondence research. Tensor networks are designed to calculate quantum mechanical systems, and MERA tensor networks, entangled quantum systems. Random tensors allow gauge theory calculations that sum over random configurations of geometries (fields). Hence, random tensor networks might be used as a new method to interrogate emergent bulk geometric structure, and to create new holographic codes as a result for the greater mobilization of the AdS/CFT correspondence to practical settings such as error correction in quantum computing.

15.2 The AdS/CFT Correspondence Generalized to the SNQFT

The benefit of the Ads/CFT correspondence is that it is both theoretical and practical (and mathematically solvable). The correspondence provides a template that can be implemented in contemporary systems such as smart network technologies. The key concept in the AdS/CFT correspondence is that there is a relationship between two domains, one that has one more dimension than the other. The correspondence is a portable and extensible analytic model.

15.2.1 *Bidirectional: Bulk–boundary linkage*

The interesting point for the SNQFT is that the correspondence can be conceived in different ways. Whereas in physics, the correspondence is employed most often in one direction, to reconstruct information about the bulk from the boundary (for example, learning about emergent structure in the bulk from the boundary), in smart network systems, there may be an extant bulk for which it is useful to abstract a boundary formulation.

The bulk–boundary relationship in the AdS/CFT correspondence can be conceived in either direction (from bulk to boundary or from boundary to bulk). There could be one region that has one fewer dimensions than the other. Or, there could be one region that has one more dimension than the other. Depending on the problem context, it is typically easier to start with the main region, which might be either the region with one fewer dimensions or one more dimension. Sometimes the detail of the bulk is known, and an abstraction to describe the overall activity is sought in one fewer dimensions. Other times, an overall metric is known, but not its constituent detail.

Complex systems may have two boundary regions and a bulk region in the middle. An economy is a complex system in this structure. A three-tier model of the economy can be elaborated as two boundaries sandwiched around a bulk. There is the macroeconomic indicator of GDP as the top boundary, a complex mass of unknown structure in the middle, and microeconomic-type indicators such as housing starts, new business registrations, and bankruptcies as the bottom boundary. With the AdS/CFT correspondence model, either boundary region might be used to interrogate the complex mass of the bulk in the middle.

Sharpening these ideas, the correspondence can be conceived as regions that are (d+1) and (d) to each other, or regions that are (d) and (d–1) to each other. The translation can go either way, from the bulk to the boundary or the boundary to the bulk (Table 15.3).

The AdS/CFT correspondence is a general tool for investigating systems that can be structured as having two regions, one with one larger dimension than the other. The correspondence is a portable technique for spanning dimensions, which is useful as many problems are in the form

Table 15.3. Examples of bulk–boundary directional transformations.

Correspondence direction			Example
Bulk	→	**Boundary**	
$d+1$		d	An existing bulk (messy reality) for which an overall description is sought
d		$d-1$	
Boundary	→	**Bulk**	
d		$d+1$	Cosmology: Find bulk locality/geometry of space and time from boundary theory
$d-1$		d	

of translating between dimensions. In one sense, the correspondence can be deployed as an information compression technique, in that a bigger dimensional region can be represented with a smaller dimensional surface. A 3D volume can be compressed into a 2D surface (i.e. a much shorter description). Going the other direction, the correspondence is an information expansion technique, allowing amplification of a level of more granular detail from an otherwise simple boundary signal.

Theory construction is a key potential use of the bulk–boundary correspondence. There are many signal-to-noise problems of needing to understand the bigger picture or emerging trend from the vast detail in the bulk. For example, abstracting pacemaker data into a predictive model for cardiac events is a contemporary challenge. A wide range of data are available, but the salient aspects of the data sequences are unknown and there is no theory to motivate the data into practical use. Likewise, the general problem of the big data era is that big data is not smart data. Having mechanisms to elicit relevant structure such as the correspondence could be useful to transform data into insights. In other cases, it is the opposite. The bulk is a black box (for example, deep learning systems, and blockchain economic activity) and hence a boundary theory might be useful to extract information about the unknown bulk at a higher-order level to produce an actionable understanding and an overall theory that describes the hidden activity.

15.2.2 *Unidirectional: Interrogate complexity with simplicity*

The AdS/CFT correspondence is a general complexity management technology (ComplexityTech) for dimensionality reduction. The bulk–boundary correspondence is flexible, in that it can be instantiated in either direction (bulk to boundary or boundary to bulk). One of the main use cases is employing the correspondence to interrogate domains of complexity (the bulk) with simpler models in one fewer dimensions (the boundary). The general principle of the correspondence is that for any complex volume, a boundary theory in one fewer dimensions can be found or defined to describe it. In fact, there are many systems that can be seen through the lens of the correspondence (Table 15.4).

The implication is that the bulk–boundary correspondence is a method that might be applied to any system through the argument that a boundary theory can describe the bulk volume in one fewer dimensions. One-way hash functions and proofs are an example. A one-way hash function is a demonstration of the bulk–boundary correspondence, in that the verification happens in one fewer dimensions than the generation. Likewise, the holographic interpretation of proofs is that since verification occurs in one

Table 15.4. Examples of bulk–boundary correspondence relationships.

Domain	Greater-dimensional bulk	One fewer dimensional boundary
One-way hash function	Hash generation	Hash verification
Proofs	Proof generation	Proof verification
Zero-knowledge proof	Prover knows the underlying data	Verifier knows one bit of information
Holographic annealing	Noisy spin glass energy optimization process	Final answer
Holographic consensus	Proof-of-work mining	Confirmed block
Blockchain Layer 2: Lightning Network, sidechains, channel factories	Multiple rounds of transactions and channel opening, closing, and rebalancing	On-chain transaction
Oil and gas futures and options trading	Speculative exchange (8–15 times/barrel)	Actual resource consumption

fewer dimensions than generation, a proof structure has a bulk–boundary correspondence relationship. A proof is a correspondence of the proof generation in the bulk and the proof verification in the boundary. The implication of proofs being in the structure of the holographic correspondence could be generating improved proof structures through the analytic models used for solving bulk–boundary correspondence problems (making more complicated, yet simultaneously efficient proof structures, per the principle of STARKs). A further application of these ideas could find an instantiation in post-quantum lattice cryptography duals, in which proofs are similarly created in one dimension and verified in another, with selectable dimensions for both bulk and boundary in non Euclidean geometries.

15.3 Adding Dynamics to the AdS/CFT Correspondence

There are at least two ways to motivate the correspondence. One is to notice that many phenomena are in the form of a bulk–boundary correspondence, in the sense of there being a relation between a bulk region and a boundary region in one fewer dimensions, and to engage the system as it naturally operates in this structure (Table 15.4). Another way is to specifically exploit and manipulate the system per its correspondence-related properties to achieve new objectives.

One idea is explicitly adding dynamics to the correspondence. By specifying the system dynamics, the AdS/CFT correspondence can be used to direct the running of messy processes in the bulk to obtain the answer as a precise signal in the boundary. The process must run, but all that is needed from the process is one answer at the end. Only the macroscale output value is needed, the "temperature" value, not all of the movements of every particle in the room. Some of these kinds of messy processes with clear endpoint signals are blockchain consensus and spin glass energy optimization problems solved with quantum annealing. A process must be run, but all that is required at the end is the salient outcome. It is not merely the output value that is required, but the output value that is produced as a result of a process that runs in time, which suggests system dynamics. The complete system consists of the process that runs in time (elapsing in time t), and the final output value from the process at the end (time t+1).

The idea is to direct the correspondence as a dynamical system that runs in time, specifying dynamics as a selectable parameter, including with different kinds of time complexity. The dynamical evolution of the correspondence appears to be a novel technophysics use of the correspondence. Although the correspondence has been used to study dynamical systems (such as the big bang, black hole evaporation, and entropy in the inflationary expansion of the universe), the correspondence as a dynamical system itself has not been explored.

The correspondence has dynamics because in the standard reconstruction of the global relationship between the bulk and the boundary regions, the boundary conformal field theory is defined as a Cauchy surface with a time dimension (Hamilton *et al.*, 2006). The higher-dimensional bulk has the complex geometric structure of time and space dimensions. The boundary likewise has a time dimension. Hence, the dynamics of the correspondence can be defined for any problem.

The bulk–boundary mapping can be established based on any variety of factors, including theories, spaces, states, operators, energy, and degrees of freedom (parameters). Dynamics can be specified for any kind of mapping. For example, holographic annealing is the idea of running an annealing process in the bulk and obtaining the answer in the boundary. The notion of operator mappings suggests Hamiltonians that measure the total energy of the system. A quantum annealer runs from the high-energy state to the low-energy state of a system. The idea is to run the annealing in the bulk region, with all of the messy details, and translate the output to a single answer on the boundary (through the bulk–boundary operator mapping). This is a holographic annealing, the idea of running a detailed process in a (d+1) dimensional region and translating the answer to concise output in one fewer dimensions in the boundary region. The system dynamics of the correspondence as a selectable property allows various holographic processes to be elaborated to unfold in the bulk to produce an answer in the boundary. Specifiable dynamics brings a new element of manipulation to the correspondence as a tool for solving problems with a bulk–boundary relationship.

15.3.1 *Spin glass interpretation of the AdS/CFT correspondence*

Interpreting the correspondence as a spin glass (as an energy optimization problem as indicated in holographic annealing) is another technophysics advance and a new way of exploiting the correspondence. The correspondence has been applied to condensed matter and superconducting materials, but not spin glasses directly. The notion is supported however, in follow-on work from Ryu and Takayanagi that formulates a holographic description of the effects of disorder in conformal field theories based on the AdS/CFT correspondence (Fujita *et al.*, 2008). Thus, it could be possible to interpret the AdS/CFT correspondence as a 3D theory in which these kinds of bulk–boundary transformations might be possible (running a process in the bulk and obtaining the answer in the boundary). Other work discusses a gauge/gravity dual, the SYK/AdS duality (Sachdev–Ye–Kitaev, SYK), as a special kind of quantum field theory used in conjunction with tensor models as a new class of large-N quantum field theories (Das *et al.*, 2017). This further supports the idea of employing the correspondence as a tool to perform manipulations between the two regions.

15.3.1.1 *Holographic consensus as a quantum error code*

Holographic annealing, and likewise holographic consensus, could operate through the bulk–boundary correspondence in a smart network use case of the AdS/DLT (blockchain) correspondence. The consensus process is a dynamical system that evolves from a high-energy state of frenetic activity (miners producing billions of nonce guesses per second) to a low-energy state (an immutable confirmed block). Similarly, the consensus could be run as a bulk process translated to a boundary answer. The dictionary mapping is essentially a quantum error correcting code, the mapping between the bulk and the boundary Hamiltonians in the anneal, and the mapping between the bulk and boundary blockchain Hamiltonians in blockchain consensus (whose output data is a confirmed block (Bitcoin) or the world state of the blockchain system (Ethereum, DFINITY)). The idea is to run the holographic processes on a quantum simulator rather than in real life, as a security and efficiency upgrade. Running blockchain consensus on a quantum annealing machine has been proposed (Kalinin & Berloff, 2018), and running a holographic consensus by specifying the dynamical properties of the correspondence extends this idea and could make it even more efficient.

Whereas the simplest bulk–boundary correspondence maps states of low-energy in the bulk to other states of low-energy in the boundary, the mapping can also include diverse mappings involving high-energy states if the code space mapping of the correspondence is expanded. A further new use case for selectable dynamics in the holographic correspondence is information lifecycle management. Two aspects are relevant, using information efficiency algorithms to suggest the right type and amount of noise to produce a certain kind of signal in a calculation, and also to manage the interval of data refresh and purge.

15.3.2 *Holographic geometry is free*

Holographic geometry is the idea of obtaining "good" (i.e. physics-based) geometry for free as part of deploying the AdS/CFT correspondence model. Another way that the correspondence might be deployed for new objectives is related to its feature of producing "good" geometry as a byproduct of its operation. The correspondence is designed to describe local structure that arises in the bulk, namely geometry. Quantum entanglement in the boundary theory is regarded as a building block for emergent bulk geometry. The idea is that geometry in the bulk arises from being describable by entanglement in the boundary theory. In the correspondence, the geometry of space and time is an emergent property related to quantum entanglement.

The benefit of using the correspondence is that geometry is obtained for free as an effect of applying the model, and so any application that requires geometry or has geometric aspects (which could be any graph, meaning any smart network system) might benefit from the correspondence. "Good" geometry is meant in the sense that physics has fundamental and well-formed geometry. Since the correspondence is physics-based, the analytic geometry obtained through the model is not some random geometry, but geometry that aligned with the physical world because it is derived from a physical theory. This is a Feynman-type argument that good quantum computation design (e.g. good quantum smart network design) is in parallel with quantum reality. Both have well-formed geometric properties. As a design requirement, quantum smart networks should be based on physics principles (a smart network is a physics-based computation network). Related work calls for a more theoretically-based instantiation of communications networks, using complexity and technophysics methods (Kelly, 2008).

The idea is to use the properties of the correspondence to obtain geometry for free since geometry is one of the bulk structures that is naturally obtained from the correspondence. The feature produced by the correspondence is holographic geometry, geometric structure in a bulk region as elaborated by entanglement in the boundary. It may even be possible to prove that holographic geometry is well-formed (by analogy to quantum statistics), that such geometry could not have been derived other than from a quantum mechanical model.

15.3.2.1 *Geometry-based smart routing*

The point is that having well-formed geometry is needed for various efficient routing activities in smart networks. In deep learning systems, holographic geometry could be used in the improved feeding-forward and backpropagating of the network processing to arrive at a solution, and also in self-analyzing the best layered architecture for the job. In blockchains, holographic geometry could be used to further develop smart routing protocols in the Lightning Network and other Layer 2 off-chain and sidechain protocols. There could be dynamic routing based on bulk geometry, as an extension to an earlier proposal for well-formed directed acyclic graph routing based on matrix formulations. Holographic geometry could be used in the instantiation of quantum payment channels (using quantum channel functionality as an ongoing conduit for resource transfer). In the quantum internet, geometry could likewise be a parameter in nonlinear multi-path smart routing.

The higher order message is that geometry matters. Already in classical networks, having technophysics-based geometry and routing could be an efficiency improvement and a complexity management technique. In quantum communications networks and quantum computing, it is clear that well-formed geometry is a crucial property. This is because qubit formulations are superpositions in Hilbert space, and gate model operations are a function of the geometric formulations of multiple qubits in vector-based coupled interactions. Since good geometry is available for free in the AdS/CFT correspondence, this feature might be exploited in many contexts. The design principle for quantum smart networks is that they too are better formed to the extent that they are in the structure of the real-life quantum world.

15.3.2.2 *Alternative geometries*

Given that geometry is an emergent property in the application of the AdS/CFT correspondence, there is an opportunity to explore alternative geometries in the AdS bulk (for example, the selectable geometries from information geometry). Riemannian geometry is indicated because the canonical theory of real-life space and time in the universe is based on Riemannian curvature. Hyperbolic geometry is also indicated because the AdS bulk is an anti-de Sitter space, a hyperbolic space that shrinks into the circle edge as in Escher Circle Limits paintings. A conjecture in geometric deep learning about the purported manifold structure of data could be tested. The claim is that high-dimensional data in its natural representation is concentrated close to low-dimensional manifolds, and that the deep learning algorithm learns this geometric structure to solve optimization problems (Lei *et al.*, 2018). Further, a property of the AdS geometry is that it is of a particular curvature scale, and hence other curvature scales might be examined, with the implication that there could be different emergence models for time and space, all tested through the AdS/CFT correspondence. A new computational complexity term could be defined to accompany time and space complexity, geometric complexity, enumerating diverse geometries as a parameter.

15.4 Quantum Information/SNQFT Correspondence

15.4.1 *Strategy: Solve any theory as a field theory in one fewer dimensions*

The key point of the AdS/CFT correspondence is the generalization that any more complicated theory can be defined as a less complicated theory in one fewer dimensions. A bulk theory can be defined as a boundary theory in one fewer dimensions. A field theory (an effective field theory) is in the bulk, and a CFT (a conformal field theory or simplified field theory in which a time dimension or other parameter must be reconstructed) is in the boundary in (d–1) dimensions.

The implication is that the bulk field theory (any theory) can be solved in a boundary theory in (d–1) dimensions. The correspondence can be used to obtain a solvable model of any system (assuming it can be sufficiently reduced and transformed with a 1D-type reduction). This is the lesson of MERA tensor networks, that even the most complex and highly-entangled quantum many-body problems may be instantiated in a tensor network model and solved analytically.

The take-away strategy is to shape any problem in the form of a field theory, and to solve it as a boundary theory in one fewer dimensions. (This is analogous to shaping any problem as a logistic regression and solving it with a deep learning optimization network.) Since many smart network problems are already in the form of a graph, it could be straightforward to rewrite them as fields, and as a field theory.

15.4.2 *Macroscale reality is the boundary to the quantum mechanical bulk*

The concept of the correspondence is that any theory can be seen as a bulk theory with a solvable boundary interpretation in one fewer dimensions. This can be extended with the idea that everyday macroscale reality is the boundary to the quantum mechanical bulk. This makes sense intuitively. It is known that macroscale objects (tables, chairs, humans) comprise atoms at the quantum mechanical scale. As such, physical reality may have a long-distance and a short-distance description of the same phenomenon (Table 15.5). For example, temperature is a long-distance description of the short-distance phenomenon of particle movement. Temperature is a macroscale instantiation of data about the movement of septillions of particles at the microscale. The macroscale is the boundary to the microscale, not literally having just one dimension removed, but one dimension removed conceptually in the generalized model of macroscale and microscale.

Table 15.5. Long-distance and short-distance descriptions in field theory systems.

Microstate environment	Bulk EFT describing microstates	Macrostate metric	Boundary CFT describing macrostates
Bulk	Effective field theory	Boundary (d–1)	Conformal field theory
Air particles	Quantum mechanics, wave function	Temperature, Pressure	Statistical mechanics
Water molecules	Particle physics, QCD	Waves	Hydrodynamics
Atoms in a crystal	Spin glass	Superconducting materials	Condensed matter physics
Quantum information structure (geometry, time, space)	SNQFT	Deep learning particle output, holographic consensus	SNFT

The idea is to use the macroscale to interrogate the quantum world through the correspondence. In this sense, macroscale reality is a boundary theory for the complexity of the quantum mechanical microscale. Just as there may be a CFT on the boundary surface of the universe that describes the bulk emergence of geometry, matter, and space and time in the universe, macroreality too can be seen as a surface theory on the quantum mechanical bulk of microreality.

15.5 The SNFT is the Boundary CFT to the Bulk Quantum Information Domain

The reason to conceive macroscale reality as a boundary theory on the quantum mechanical bulk domain is to use this construction to realize the smart network theories. The SNFT is the boundary CFT with which to instantiate the structure of the quantum realm, which is described by the SNQFT. Just as boundary entanglement entropy is used to elaborate emergent geometrical structure in the bulk as a minimal surface (per the Ryu–Takayanagi entanglement entropy formula), so too is the SNFT employed as a boundary CFT with which to define structure in the quantum information theory domain. The idea is to enumerate bulk quantum information theoretic structure (related to geometry, space, and time) from the corresponding boundary CFT, which is the SNFT. Macroscale reality is one fewer dimensions in time or space in terms of quantum computational complexity. Optimal quantum information structures are unknown, but might be calculated with the SNFT.

One such quantum mechanical bulk to which the AdS/smart network correspondence might be applied is high-energy physics. The next upgrade of the LHC, the High-Luminosity Large Hadron Collider (HL-LHC), is expected to begin operation in 2026 with an estimated required computing capacity of 50–100 times greater than what currently exists (Carminati, 2018). Quantum computing is implicated, and smart network field theories to interpret the regime.

15.5.1 *The internet as a quantum computer*

The internet as a quantum computer could be an analytic tool (a tensor network) for the realization of the SNFT in calculating the emergent bulk quantum information structure. The SNFT is the boundary CFT with which to calculate the emergent bulk structure, in this case, quantum information theoretic structure such as geometric, spatial, and temporal computing paradigms. The internet as a quantum computer provides an analytic model for calculating the AdS/CFT correspondence directly. (Other methods could also be used such as quantum error-correction code schemes implemented by defining mutual bulk–boundary subspaces connected with subalgebras and entanglement wedge reconstructions.)

There is a precedent for the idea of the internet as a quantum computer. Given their global reach, scale, complexity, and particle-many elements, the internet and the quantum internet have been suggested for applications beyond their primary use in communications and data transfer. One such meta-use is the internet as a scientific laboratory. Deploying the quantum internet as a science laboratory has been proposed, in particular for precision sensing that might be used to detect gravitational waves (Castelvecchi, 2018). The point is that (as illuminated in the smart network concept), smart networks are computation networks (with executable code). The network itself is the computer (a point which Ethereum demonstrates). The whole internet itself is the computer, perhaps ultimately a quantum computer.

15.5.2 *Computing particle-many systems with the quantum internet*

The internet as a quantum computer could be used to compute particle-many systems. One reason for instantiating the internet as a quantum computer is to compute the emergent bulk structure, meaning useful theories with which to further instantiate quantum information models that have interesting geometric, temporal, and spatial properties. The other reason is to calculate more about the boundary (i.e. macroreality). Macroscale reality is a quantum computable domain with particle-many systems. Before attempting to understand quantum systems directly, an intuition is to study macroscale systems that are somewhat in the form of many-particle quantum systems.

Macroscale particle-many systems include economies, food webs, ecological models, energy grids, social networks, and transportation networks. The brain, the human body, and biology, in general, are particle-many systems. Avogadro's number, a number so big that it usually only appears in biology and chemistry, is starting to surface in the context of quantum computing. Avogadro's number is about a trillion times a trillion (more specifically (6×10^{23}) or (0.6 of a trillion × a trillion)). A quantum computer with 79 entangled qubits has an Avogadro number of states (since with quantum entanglement, n qubits can represent $2n$ different states on which the same calculation can be performed simultaneously).

The intuition is to identify the largest existing particle-many domains at the macroscale, instantiate them with quantum computable properties (superposition, entanglement, interference), and run the calculations. The premise is to use the macroscale boundary to learn about the microscale world. The assumption is that macroscale reality provides an ideal laboratory for studying quantum mechanics. Macroscale reality is a platform for instantiating bulk problems in one fewer dimensions and rendering them solvable. SNFT is the boundary CFT with which to interrogate the bulk EFT, which is the SNQFT.

15.5.2.1 *Macroscale reality as a physics laboratory*

The endgame of smart network theories is not just the development of better smart networks that is merely the proximate objective. One long-term objective of smart network theories is to instantiate the entire edifice of macroscale reality as a giant physics lab. Many scientists already use macroscale reality as a lab, not usually the whole thing, perhaps only because methods (smart network theories) and tools (quantum computing) have not yet been available. However, now it is increasingly possible to employ the whole of macroscale reality as the lab or data set. LIGO and the LHC are expensive artificial scientific labs for specialized purposes, a supplemental idea is to harness macroscale reality. For some studies, the lab has to be a living lab precisely because it is a well-formed physical system whose requisite complexity could not have been simulated. Since macroscale reality is a lab generated by real-life phenomena, it is a math-obeying model with physical properties conducive to the correspondence and quantum computable possibilities. Macroscale reality is a long-distance description of microscale physics, it is the boundary to the quantum mechanical bulk, and as such, could be a solvable model for the bulk in one fewer dimensions.

15.6 Risks and Limitations

There could be many risks and limitations to the technophysics approach and smart network theories (SNFT and SNQFT) developed in this work. One class of risks concerns the underlying technologies to which the approach and theories pertain. The technologies, namely quantum computing, blockchain, and deep learning, are still in early stages of development with considerable uncertainty as to their evolution. One complaint could be that it is simply too early to postulate theories that may become hopelessly incorrect or outdated, especially those related to the quantum domain. However, the counterargument is that it is never too early for theory development, and the kind of work the theory should do in terms of specifying the scope for the explanatory mechanism, the constituent elements, and the predictions that can be made, precisely in order to provide guidance in situations of uncertainty. Any scientific discovery process is always open to revision, including the theories that support it.

Another class of risks pertains to the theory development. A critique could be levied that it is difficult and possibly inaccurate to develop theories when there is a lack of scientific agreement about many of the underlying concepts and proposals. For example, not all scientists agree that there can be information-theoretic interpretations of physical phenomena. The holographic principle is not necessarily widely accepted, much less its potential reach in extended applications. Also, there is a diversity of ideas about the extent to which quantum theories and geometry theories may be blended. However, science is never a settled matter, and the point of this work is to engage emerging concepts from active research frontiers of scientific efforts in many fields. The claim of this work is not that the underlying theories are accurate in their domain or beyond, but rather that they might be a useful model to interpret and apply in other domains. There could be a complaint of overreaching, that the underlying concepts might only appear to apply to smart networks when in fact there is no validity to this approach. Further, the theories fail to take up ethical technology design, computational ethics standards, and the potential impact of smart technologies on humanity. Catastrophic risks to network technologies such as electromagnetic pulses (EMPs), whether malicious or unintentional, are not discussed. These are crucial issues, but not addressed within the scope of the current work. The aim of this work is to articulate foundational physically-based theories as the strongest potential candidate for explanation in the smart network domain.

A third class of risks pertains to theory application. Perhaps when the smart network theories are applied literally, they are too focused to be of use, and when applied analogically, are too vague to find a grasp. However, the point of a theory is that it is a tool that can be tested as to the range and validity of its application. A critique could be raised that even if applied, smart network theories do not necessarily guarantee a useful resolution to problems. The smart network theories lack more detailed demonstration examples and experimental results. They might fail to make predictions that are in alignment with actual physical phenomena, and thus do not offer a practical benefit. The smart network theories are a complexity technology, designed for application to complex systems, but criticality and phase transition are notoriously difficult to capture with analytic models. A model is only a representation of the territory it attempts to map. Aspects may arise in the underlying system that are outside of the model's frame. However, exactly because smart networks are complex systems whose behavior might change and evolve in unexpected ways, theoretically-based methods for studying and managing them could be useful. The methods proposed by smart network theories could lead to a more rigorous study of complex systems.

15.7 Conclusion

This work uses a technophysics approach to propose smart network theories (SNFT and SNQFT) which could provide the start towards a comprehensive conceptualization of smart networks as a novel computational emergence, and indicate how they may play a role in the development of quantum information systems. The computability of increasingly larger and more complex domains is a central theme, and a causal model underlying the progression of smart networks is developed during the course of the book (Figure 15.1). The hypothesis is that computation capability is the biggest current factor in acquiring knowledge about physical reality at all scales from the observable universe to the Planck scale. Computation capability is likely moderated by factors such as quantum properties, smart network theories, and quantum security features (the tools and apparatus by which the capacity for efficient computation is deployed towards the end of an improved understanding of physical reality).

15.7.1 *From probability to correspondence*

As the granular comprehension of physical reality grows, models for its understanding are paramount. The AdS/CFT correspondence encapsulates the current moment of understanding. Probability has been a central organizing principle for the understanding of reality, from quantum mechanics

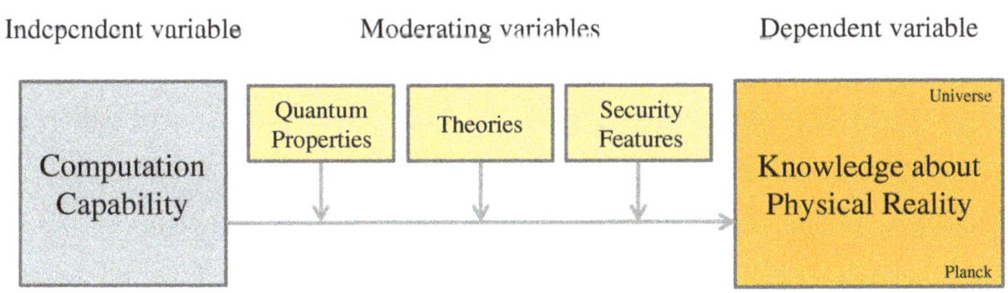

Figure 15.1. Model of computational reality with moderating variables.

to smart network frontiers (zero-knowledge proof technology and deep learning adversarial networks). Now, though, the correspondence could be a new central organizing principle for certain ways of understanding physical reality and the creation of systems to further engage it (such as atomically-precise materials and manufacturing systems), it provides the ability to portably scale diverse dimensions and solve them analytically. The correspondence also emphasizes the connection between the quantum and non-quantum domains such that the quantum realm can be activated in a useful way at the macroscale. A conceptual step beyond probability is correspondence. Whereas contemporary smart networks are state machines that make probabilistic guesses about reality states of the world and act automatically on this basis, future smart network technologies might act on the basis of the correspondence, constantly and portably solving higher-dimensional problems in surface theories.

A tool connotes factors related to its use, a climate of related assumptions that constitute the enabling background for the tool. A hammer connotes a nail. Probability connotes many factors beyond the immediate execution of a calculation. Probability includes computational complexity, statistical distributions, mathematical equations, optimization, the concept of chance, and similar formulations in both the natural world and algorithmic models (e.g. spin glasses, a fair coin toss, machine learning). The AdS/CFT correspondence is a level beyond that, a particularly rich juggernaut formulation that presumes many other concepts and mobilizations. The AdS/CFT correspondence includes the holographic principle, 3D bulks represented in 2D boundary surfaces, reconstruction, information compression, entanglement, dynamics, system criticality, emergent structure, security features, computational complexity, gauge theory and gravity, bulk and boundary linkage, and the use of dictionary mappings and codes to translate between the two regions. Considering probability and correspondence is an apples-to-oranges comparison because the very concept of the AdS/CFT correspondence denotes such a rich consolidation of functionality.

15.7.2 *Farther consequences: Quantum computing eras*

The potential realization of universal fault-tolerant quantum computers is just one next step. There could be many eras of quantum computing in the

Table 15.6. Eras in quantum computing, physical theory, and smart networks.

Quantum computing eras	Scale	Physical theory	Smart network theory
Pre-quantum Computing	10^1 m	Newtonian mechanics	Smart network field theory
Quantum Computing 1.0	10^{-9} m	Quantum mechanics	Smart network quantum field theory
Quantum Computing 2.0	10^{-15} m	QCD/gauge theories	Smart network gauge theory
Quantum Computing 3.0	10^{-35} m	Planck scale, spin networks	Smart network spin theory

farther future (Table 15.6). The first level of quantum computing (Quantum Computing 1.0) invokes basic quantum mechanical systems, attempting to harness the movement of atoms and photons for computational processes. The next level down in scale (Quantum Computing 2.0) could engage quantum chromodynamical models of subatomic particles such as gluons and quarks (with a key objective of deriving a theory of quarks, which is not yet extant). The next scale level down (Quantum Computing 3.0) could target the smallest known level of reality, the Planck scale, and its potential spin network composition, describing the smallest fundamental "atoms of reality". Spin networks are a snapshot of reality at the Planck scale, and spinfoams contain their history. Notably, spinfoams can be interpreted as either a Feynman diagram (a quantum mechanical view) or as a lattice construction (a geometric view). There are already spin network formulations of the AdS/CFT correspondence (Bodendorfer, 2016). Hence, one key principle of the correspondence (gauge/gravity duality) persists at the Planck scale, which is being able to see a system in complementary views as both quantum mechanical and geometric. The same kinds of principles might be directed towards scaling up systems to understand very-large domains (such as Earth 10^7 m, Milky Way 10^{21} m, and the observable universe 10^{27} m) in addition to those of the very small.

References

Almheiri, A., Dong, X. & Harlow, D. (2015). Bulk Locality and Quantum Error Correction in AdS/CFT. *J. High Energ. Phys.* 163, 1–33.

Baratin, A. & Oriti, D. (2010). Group field theory with non-commutative metric variables. *Phys. Rev. Lett.* 105(221302).

Bodendorfer, N. (2016). A note on conformally compactified connection dynamics tailored for anti-de Sitter space. *Class. Quantum Grav.* 33(237002).

Carminati, F. (2018). *Quantum Thinking Required.* Cern Courier.

Castelvecchi, D. (2018). *Here's What the Quantum Internet Has in Store.* Scientific American.

Das, S.R., Jevicki, A. & Suzuki, K. (2017). Three Dimensional view of the SYK/AdS duality. *J. High Energ. Phys.* 09:017.

Fonseca, A., Rosier, A., Vertesi, T. *et al.* (2018). Survey on the Bell nonlocality of a pair of entangled qudits. *Phys. Rev. A* 98:042105.

Fujita, M., Hikida, Y., Ryu, S. & Takayanagi, T. (2008). Disordered systems and the replica method in AdS/CFT. *J. High Energ. Phys.* 0812:065.

Gurau, R.G. (2016). *Random Tensors.* Oxford UK: Oxford University Press.

Hamilton, A., Kabat, D., Lifschytz, G. & Lowe, D.A. (2006). Local bulk operators in AdS/CFT: A boundary view of horizons and locality. *Phys. Rev. D* 73: 086003.

Hayden, P., Nezami, S., Qi, X.L. *et al.* (2016). Holographic duality from random tensor networks. *J. High Energ. Phys.* 11:009.

Kalinin, K.P. & Berloff, N.G. (2018). Blockchain platform with proof-of-work based on analog Hamiltonian optimisers. arXiv:1802.10091 [quant-ph].

Kelly, F. (2008). The mathematics of traffic in networks. In: Gowers, T., Barrow-Green, J., and Leader, I. (eds). *The Princeton Companion to Mathematics.* Princeton University Press, pp. 862–70.

Lei, N., Luo, Z., Yau, S.-T. & Gu, D.X. (2018). Geometric understanding of deep learning. arXiv:1805.10451 [cs.LG].

Watrous, J. (2002). Quantum statistical zero-knowledge. arXiv:quant-ph/0202111.

Glossary

AdS/CFT correspondence: Anti-de Sitter space/conformal field theory (AdS/CFT) correspondence (also called gauge/gravity duality and the bulk/boundary correspondence) is the proposed correspondence between a volume of space and its boundary region such that the interior bulk region can be described by a boundary theory in one fewer dimensions.

Artificial intelligence (AI): Artificial intelligence is using computers to do cognitive work (mental and physical) that usually requires a human.

Blockchain (distributed ledger) technology: A blockchain is a distributed data structure that is an immutable, cryptographic, consensus-driven ledger.

Computational complexity: Computational complexity is the computational resources in terms of time and space (classical or quantum) that are necessary to calculate a given problem.

ConsensusTech (Consensus technology): Consensus technology is technology used for the self-coordinated agreement and governance of any multi-agent system (human or machine).

Deep learning: Deep learning is a class of machine learning algorithms in the form of a neural network that uses a cascade of layers of processing units to model high-level abstractions in data and extract features with which to make predictive guesses about new data.

Deep learning chains: Deep learning chains is a control technology for fleet-many internet-connected items with object recognition (deep learning) and secure tracking (blockchain).

Deep learning proofs: Deep learning proofs refers to a deep learning system in which perceptrons generate proofs of their node's contribution.

Entropy: Entropy is a measure of the number of total possible configurations of a system.

Field: A field is (1) the precise physical definition of an electromagnetic or gravitational field; (2) the ability to control fleet-many items as one unit; (3) a function with a value at every location (also a graph or matrix).

Hamiltonian term: A Hamiltonian term or operator is (1) a function that provides a mathematical description of a physical system in terms of energy; (2) a function that produces a point value corresponding to an underlying dynamical system configuration.

Hash code: A hash code is a fixed-length function output used to map data of arbitrary size onto data of a fixed size.

High-dimensionality: High-dimensionality refers to a data set with more features than samples.

Holographic code: A holographic code is a mechanism for mapping bulk and boundary regions in the AdS/CFT correspondence.

Holographic principle: The holographic principle is the notion of reconstructing a 3D volume on a 2D surface (similar to a hologram). The upshot is that there are two valid views of the same physical situation.

IDtech (identification technology): IDtech is the ability to recognize objects (whether physical or digital).

Machine learning: Machine learning is a statistical method in which computers perform tasks by relying on information patterns and inference as opposed to explicit instructions.

Merkle tree: A Merkle tree is a hierarchical structure of hash codes corresponding to a large data structure. A hash is made for each data element, then a hash of these hashes, and so on, hierarchically up until there is just one top-level hash for the data structure, the Merkle root.

NISQ device: A NISQ (noisy intermediate-scale quantum) device is a near-term quantum computer that is able to conduct some degree of quantum information processing despite imperfect physical components.

Quantum error correction: Quantum error correction is the technique of smearing out the information of one quantum information bit (qubit) onto entangled qubits such that it can be recovered if damaged.

Quantum smart networks: Quantum smart networks are smart network technologies implemented at the quantum scale (1×10^{-9}m) or in quantum computing environments.

Renormalization: Renormalization is a mathematical technique that allows a system to be investigated at different scales, especially as it changes dynamically.

Quantum internet: The quantum internet is a very-fast ultra-secure future internet concept based on technologies such as quantum key distribution, entanglement, and quantum memory.

Smart networks: Smart networks are intelligent self-operating computation networks such as blockchains and deep learning neural nets.

Smart network field theory: A smart network field theory (classical or quantum) is a field theory for the orchestration of particle-many systems from a characterization, control, criticality, and novelty emergence perspective.

Spin glass: A spin glass is a disordered magnet that is a metastable system in which roughly half of its molecular bonds are spin-up, and half spin-down, which can be used as a computation system.

Technophysics: Technophysics is the application of physics principles to the study of technology.

Temperature term: A temperature term is an aggregate informational state of a system that might be employed as a control lever.

Time complexity: Time complexity is the computational complexity related to the time required to calculate a certain problem. (By analogy, geometric complexity connotes diverse geometric computational classes based on the underlying geometry used in the problem.)

Web 3.0: Web 3.0 is a functionality, collaboration, and trust upgrade for the internet's operating infrastructure (Web 1.0 was the read web, Web 2.0 the read/write web, and Web 3.0 the read/write/trust web), including new standards such as hash-linked data structures to call data uniformly.

Zero-knowledge proof: A zero-knowledge proof is a proof that reveals no information except the correctness of the statement. Data verification is separated from the data itself, conveying zero knowledge about the underlying data, thereby keeping it private.

Index